Nuclear Power and Social Planning

Written under the auspices of the
Center of International Studies, Princeton University

A list of other Center publications
appears at the back of this book

Nuclear Power and Social Planning

The City of the Second Sun

Gerald Garvey
Center of International Studies
Princeton University

Lexington Books
D.C. Heath and Company
Lexington, Massachusetts
Toronto

Library of Congress Cataloging in Publication Data

Garvey, Gerald, 1935-
Nuclear power and social planning.

Bibliography: p.
Includes index.
1. Atomic power industry–United States. 2. United States–Social policy. I. Title.
HD9698.U52G37 309.2'12 76-54556
ISBN 0-669-01303-x

Published simultaneously in Canada.

Printed in the United States of America.

International Standard Book Number: 0-669-01303-x

Library of Congress Catalog Card Number: 76-54556

For ~~Sarah~~

LESLIE

Contents

List of Figures

List of Tables

Preface

Analyses of America's energy policies may rest on either of two assumptions. Some analyses set forth from the assumption that policymakers have virtually unconstrained opportunities to choose the course of the future. Others assume that certain past commitments almost irresistibly commit the nation to continued reliance on particular energy forms. This book falls in the latter category.

America's commitment to nuclear power, I believe, is irreversible. The central issue of American energy policy, then, is not *whether* to move forward with the atom, but rather *how best to control nuclear development.* This issue in turn may be approached in terms of four interrelated questions:

1. Should the United States continue to develop electric generating plants based on the fissioning of uranium, and perhaps eventually of plutonium, in light-water reactors?
2. Can an expanding nuclear base in the years 1980 to 2000 be premised on hopes for a switchover to controlled thermonuclear power in the early twenty-first century?
3. Should the country's nuclear capacity, whether based on fission or on fusion, be sited in several hundred widely dispersed small-sized stations, or in 25 to 30 concentrated facilities called *power parks*?
4. May power parks, if built, be specifically designed to stimulate economic development on America's sparsely populated intermetropolitan frontier?

The technical, economic, and sociopolitical factors that condition the answers to these four questions define the subject of the following pages.

This book is intended as more than a concrete proposal for the direction of American nuclear policy. It is intended as an intellectual provocation to students in a variety of college-level courses—courses in energy and environmental studies, in "science and public affairs" and in the policy sciences generally, in public administration, in regional planning, and in economic development.

The exposition responds to a basic defect—a narrowness of perspective—in the way Americans on both sides of the nuclear dialogue are setting the terms of future energy development. Electric utility executives, officials of energy-oriented government agencies, utility lawyers, and environmentalists often focus on the technical and commercial aspects of energy supply. Such a limited vision virtually guarantees that the full range of potential costs and benefits of particular technologies will fail to be recognized. A main point of this book is to establish the intimate connection that potentially exists between energy policy and social planning.

Since the issues themselves are controversial, the approach taken in these chapters can hardly but contribute to the heat of the national debate on nuclear power. Representatives of all opinions have something to contribute to the ultimate resolution of the surpassingly complex issues of nuclear policy. I owe much to many, and not least to those whose conclusions on nuclear power differ from my own. "Certitude is not the test of certainty," Oliver Wendell Holmes once observed. The good humor and patience with which antinuclear spokesmen have assisted me underscores the pertinence of Holmes's remark to a field in which the last word will not be heard for many, many years.

In this connection I would enter a special word of thanks to the following Princeton faculty members who read all or parts of the manuscript: Robert C. Axtmann (Chemical Engineering); Duane Lockard (Politics); Robert G. Mills (Plasma Physics Laboratory); Frank von Hippel (Center for Environmental Studies); Kenneth Rosen (Economics). Ms. Janet Lowenthal of the Princeton Citizens for Responsible Power Policies also went over the entire manuscript and provided valuable insights into the views of basically antinuclear partisans.

To no one must I avow a greater professional and personal debt than to Mr. Robert Brenner, currently a graduate student at the Princeton University Woodrow Wilson School. His senior thesis prepared under my supervision, *Nuclear Energy Parks and Regional Development,* represented at the time of its completion in April 1975 a most thorough consideration of the energy park concept, embedded in a sophisticated economic analysis of costs and advantages. Rob has remained a source of counsel, expertise, and most highly valued friendship.

I could scarcely overstate my debt to the members of two official panels on which I served as a regular participant during the period in which the idea of the city of the "second sun" began to take shape.

The first, the National Academy of Public Administration (NAPA) Panel on Energy Parks, William E. Warne, Chairman, convened in Washington, D.C. in August 1974. This group, which has met regularly and remains a working panel of the National Academy as of this writing (November 1976), included Messrs. James Baroff, Marc Messing, James Sperling, James Sundquist, Joseph Swidler, Arvin Upton, and G.O. Wessenauer. Other regular participants in the activities of the NAPA panel during the period of the General Electric study cited in the text were Harry Finger, Joseph Leiberman, and Anthony Mione of General Electric; E. David Doane of LeBoeff, Lamb, Lieby and MacRae; and Erasmus Kloman of the National Academy.

Second is the Panel on Community and Labor Impacts of Nuclear Energy Centers, which met for three days of intensive deliberations in early May 1975 in Portsmouth, New Hampshire as part of the Nuclear Regulatory Commission Nuclear Energy Center Site Survey. This panel included Messrs. George Bugliarello, James Baroff, James Cline, Harry Finger, and Willis Hay. Further substantive inputs in connection with this study effort were provided by members of the NRC Office of Special Studies: Seymour Smiley, Malcolm Ernst, Robert Jaske, William Kirwan, Stephen Salomon, and George Sege.

The typing staff of Princeton's Center of International Studies under Mrs. Jean MacDowell—and especially Gail Lubetsky—contributed far above and beyond the normal call.

Finally, I would mention, with love and appreciation, my daughter Sarah, who has helped in all the ways that a seven-year-old possibly can—and for whom, in more senses than one, I wrote the book to begin with.

Introduction: Nuclear Power and Public Policy

Einstein's equation $e = mc^2$ established the theoretical basis for nuclear power. This equation asserts the ultimate equivalence of energy to mass times the speed of light squared. So if the quantity of reacting mass in any physical process can be reduced, the loss of mass must show up as a release of energy, ultimately in the form of heat.

Such reductions can occur through two processes. In *fission*, the nucleus of a heavy element such as uranium is bombarded by a subatomic particle, a neutron. The bombarded nucleus breaks into smaller products. The difference between the binding energy of the uranium nucleus and the binding energy of the several fission products measures the energy released as a result of the breakup. By contrast, *fusion* consists in a joining of two extremely light elements. The neutrons and protons in the nucleus of a newly formed helium atom need less energy to bind them together than do the structures represented by the two hydrogen nuclei which must combine to form helium. Again, the difference appears as released "nuclear energy."

From the first, commentators have referred to nuclear reactors as models of the sun. Fission reactors may properly be referred to as metaphors for the sun. Fusion reactors, when perfected, will literally be "second suns." They will produce heat through exactly the same kinds of nuclear reactions that account for the solar flux itself.

In the late 1940s and throughout the 1950s, the promise of atomic power led to the investment of billions of dollars of public funds in the development of nuclear fission for peaceful purposes. By the late 1950s, commercial reactor developers and electric utility companies had entered the nuclear field. The ensuing decade saw a dramatic surge in orders for new nuclear electric plants. But costs mounted through the 1960s; falling uranium reserve levels began to prompt doubts about reactor fuel availability; and critics questioned the safety of nuclear plants. It became clear that the late 1970s and the 1980s would be years of restudy, reappraisal, and perhaps even of retrenchment within the field of American civilian nuclear policy.

What would be the elements in a soundly conceived policy for the future of atomic power in the United States? A responsible nuclear policy must be: (1) evaluative, (2) multidisciplinary, (3) comprehensive, (4) future-oriented, and (5) political. It would be helpful to consider each of these criteria in turn.

Policy-oriented analysis looks to the application of knowledge in order to achieve specific objectives. These objectives are implicitly identified with improvements in the quality of life within society. Thus the goals toward which a policy aims reflect a deeper and broader set of values. It is these values that the formulator of public policy must make explicit. In the following chapters, particular lines of nuclear development are advocated not to ensure electric power for the sake of having more electric power. They are advocated not merely to ensure the economic health of the multibillion-dollar nuclear industry. Rather, they are advocated because U.S. energy sufficiency seems consistent with the more basic value of national economic growth. Underlying the case for the atom is the case for economic growth.

Conversely, the most powerful line of attack against a given policy takes the form of rejection of the more basic social values that the policy in question would advance. For this reason, the first requisite of policy is that it be *evaluative* in the sense of candidly revealing the relationship between the concrete objectives of a given program and the deeper issues that divide society.

A proper evaluation of a specific policy—such as a policy aimed at rapid development of fusion power to replace fission, or a move to site all nuclear plants in America's more sparsely populated rural areas—requires a spacious view of the interaction between technology and society. The ascendancy of the "expert" can lead to a dangerous narrowing of perspective as problems are parceled out to more or less exclusive intellectual jurisdictions. The complexities of nuclear issues suggest the inadequacy of fragmentary approaches.

Questions of atomic development always hinge on inputs from physical scientists (including biologists and public health specialists), engineers, and economists. Hence Chapters 1 to 4 focus on technological issues. But responsible policy formulation must also enlist contributors from other disciplines. Demographers, urban planners, and authorities on regional development have major roles to play. And, as we shall see in Chapter 5, a proposal within the nuclear industry to gather dozens of reactors into huge nuclear complexes has brought seismologists and atmospheric scientists to a position of unexampled prominence in the nuclear field. Hence the second requisite of nuclear policy—that it be *multidisciplinary,* despite the conflicts of intellectual priorities and the inconsistencies of intellectual position that appear when melding diverse academic specializations into a single case study.

The third requisite of nuclear policy follows as a corollary of the second. The nuclear field is filled with uncertainties and risks. The cross play of conflicting interdisciplinary perspectives invariably brings a wider range both of known factors and of uncertain contingencies into the analysis. Practitioners of different academic disciplines typically handle unknown variables in different ways. An economist might try to devise means to "buy insurance" against risks, so that an adverse event could at least return some pecuniary compensation to the affected individual. By contrast, a public health official might categorically reject even insured risktaking if the adverse event—say, an accidental release of radioactivity—might mean the induction of thousands of cancers.

Ignorance of factors and contingencies which play no role in an analysis from the physicist's point of view, or the economist's, or the ecologist's, may simplify policy analysis. But such simplification is fraudulent and irresponsible. Hence the third requisite of nuclear policy is that analysis be *comprehensive.* It must reckon with, even if it cannot fully solve, the major issues and reservations suggested by a multidisciplinary approach. And it must embrace an awareness that some inconsistencies will probably remain irreconcilable and some uncertainties, irreducible.

The criterion of comprehensiveness inspires to an attitude of open-mindedness, whereby policy analysis can become a continuous, cumulative process. In this spirit, the analysis that follows represents not a final set of answers but an array of first premises and judgments.

How should the known factors and the uncertain contingencies that enter into nuclear policymaking be weighed against one another? Both "knowns" and "unknowns" must be projected decades into the future. The high costs of reactors are known factors in the late 1970s, but figures become increasingly speculative for the 1980s and 1990s. Other major variables which are now "unknowns" will become "knowns" before the turn of the century. For example, the range of U.S. population variance in the year 2025 will depend on the birth rate in the 1980s and 1990s. Because nuclear plants are not only expensive but long-lived (normally rated at 35-year useful lives or more), the new units of 1980 will influence lives in the early twenty-first century. Future population levels will decisively affect demand, and future reactor costs will affect the financing of units brought into service in the 1990s to furnish electricity for the citizens of the year 2025.

The fourth requisite—that nuclear policy be *future-oriented*—implies an awareness of the need to plan over the decades rather than merely over the years. This awareness should serve to correct the single most glaring error of omission in America's nuclear dialogue: inattention to the relationship of fission power to controlled thermonuclear fusion power.

Within the next generation, Americans may have the opportunity to begin switching from fission to fusion. Whether this opportunity will present itself depends in large part on decisions to be made within the next decade—especially on decisions regarding the funding of fusion research and the extent to which fission development between now and the year 2000 should be permitted to preempt the potential fusion market of the more distant future. As we shall see, these issues are crucial from the standpoints of nuclear economics and nuclear safety. And they are absolutely crucial from the standpoint of nuclear risk-taking, since controlled fusion power is not yet a proven technology. The entire nuclear dialogue should be recast into a multidecade time frame, with explicit consideration of prospects for a transition from fission power to a national electric system based on fusion.

Because uncertainties and risks are inherent in the nuclear technology, the field of nuclear policy is loaded with intellectual landmines. The policymaker who adopts a multidisciplinary, comprehensive, and future-oriented analysis can anticipate some of these landmines and can avoid others.

But risks of major proportions will remain. And they will remain whether America moves toward fusion power (which may not even prove commercially feasible), toward intensified fission development (in the face of unsolved problems of radioactive waste disposal), or toward an outright deemphasis of nuclear power (despite the dangers of inadequate energy supplies should solar development flag and coal development prove environmentally unacceptable). The American public at large will ultimately bear the risks inherent in any line of nuclear policy. Hence the terms of a policy proposal should contribute to the public dialogue on nuclear power.

The fifth requisite of an approach to nuclear policy, then, is that it be *political*—not in a partisan sense, but in the sense of helping to reclaim decisions from the preserves of experts and special pleaders. Policy proposals should be couched in plain enough terms to stimulate open discussion of the options—and the uncertainties and risks—in the public forum. In the same vein, new institutions of democratic choice may be needed to promote the open-endedness of the nuclear debate and to provide mechanisms for continuous learning by the public generally, as well as by professional analysts. Chapters 7 and 8 stress the importance of political forums at the "forgotten level" of American government—the regional level—in which nuclear issues may gain thorough, informed debate.

An integrated set of programs to meet these five criteria would result from a difficult choice among several possible courses of action. The elements of such an integrated set of programs follow:

1. Minimize the buildup of nuclear fission plants, subject to stiffened safety and radiation emission standards, with stringently enforced safety rules for the reactors that American manufacturers sell abroad.
2. Bypass the breeder reactor, thereby reducing reliance on the ultrahazardous radioactive element, plutonium.
3. Accelerate efforts to develop fusion power, aimed at a transition away from fission early in the twenty-first century.
4. Adopt innovative techniques for controlling fission power during the decades that would intervene before the switchover to fusion, as in the siting of nuclear plants in specially secured clusters known as *power parks*;
5. Use power parks as growth poles to stimulate economic development in regions now suffering from stagnation and decline.
6. Delegate significant responsibilities for nuclear policy formulation to regional (multicounty and multistate) institutions that would relate energy development to the socioeconomic needs of their respective areas.
7. Encourage, where the regional authorities approve, the siting of new agricultural and industrial enterprises near power parks to promote the growth of economically dynamic "regional cities"—cities of the second sun.

The chapters that follow divide of themselves into two parts. Part I, embracing Chapters 1 to 4, deals with the technical and economic aspects of nuclear technology that provide essential grounding for policy formulation. Of particular significance are the cost implications of nuclear power; reactor safety, materials safeguards, and radioactive waste management; and the relative merits of fission and fusion.

Part II, embracing Chapters 5 to 8, applies the conclusions of the prior chapters in a broader socioeconomic setting. Power parks are considered as institutional responses to some of the most vexing economic and safety-related problems of fission reactors. Power parks are appraised for their bearing on the chance of a switchover to fusion and for their potential to serve as economic growth poles. As we shall see, these two issues are closely interrelated. The growth pole discussion is then extended to the issue of regional economic stimulation, and thence to the promotion of related agroindustrial development to underwrite the growth of new regional cities on America's "intermetropolitan frontier."

Part I
The Atom in Your Future

Introduction to Part I

The first generation of uranium-burning electric plants may be thought of as extending from the start-up in 1957 of the 90-megawatt station at Shippingport, Pennsylvania until the early 1980s, when fissionable plutonium could begin entering the industry's fuel inventory in significant quantities. Figure I-1 shows in schematic form a few of the leading reactor types, together with the actual or estimated timing of their introduction as commercial energy sources.

The second generation might stretch from the introduction of plutonium through a period of reliance on breeder reactors. Breeders, which produce plutonium for recycling as reactor fuel, may be based either on an elaboration of fission devices or on hybrid systems that link fission and fusion reactors. However, the dangers of plutonium—a long-lived element named after the god of Hades—suggest the desirability of a complete breeder bypass in order to minimize the stores of this toxic substance in America's fuel inventory. A breeder bypass would imply a stretch-out of the first generation of reactors, leading directly to the introduction of controlled fusion power early in the twenty-first century.

Fusion power, though known to be possible in principle, may not be proved feasible until sometime in the 1980s. Conceivably, a successful laboratory experiment might be delayed until the 1990s or even later. But as we shall see, progress to date by engineers and physicists suggests that thermonuclear fusion can open an era of virtually inexhaustible power, probably at a net cost advantage compared to fission-based electricity, and with a decided improvement in terms of environmental impact.

A transition from predominately coal-fired to fission-based electricity before the turn of the century, and then from fission to fusion, would describe a truly Promethean progression—so-called after the mythical figure who stole fire from the gods and gave its power to men.

Nuclear energy, whether in the form of fission or fusion, is not without formidable problems. But the problems of potential competitor forms seem even more severe—severe enough to raise doubts about the ability of oil or gas, coal, solar power, or any of the exotic energy sources (such as oil from Western sands and tars) to answer America's fuel needs.

Since 1973 petroleum prices have skyrocketed. Oil- and gas-fired electric generation will become prohibitively expensive as reservoirs, both domestic and foreign, near depletion.

Coal-fired electric plants pose more severe pollutive health hazards than do nuclear plants. This also does not take into account the environmental degradation that would be the result of strip-mining the Great Plains' coal fields, or the thousands of miners likely to suffer maiming in the deep mines of a coal-fired economy. Cost estimates become notoriously imprecise as one projects further

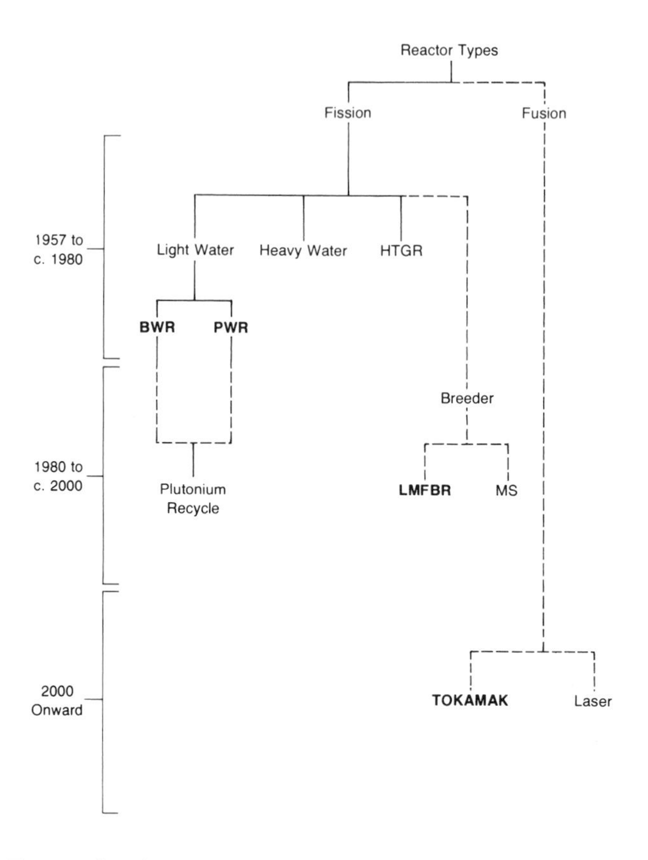

Figure I-1. The Generations of Nuclear Electric Technology.

Notes to Figure I-1. Solid lines indicate technologies that are already in use, or else the approximate year in which the indicated reactor types will become operational. Dashed lines indicate lines of evolution in nuclear technology that could produce additional successful reactors.

Bold type indicates the probable "dominant species" within each general category of reactor types. Thus while the "molten-salt" breeder appears as a realistic possibility, the liquid-metal fast breeder seems the more likely primary type.

Light-water reactors (of both the boiling- and the pressurized-water varieties) dominate the first generation technologies of American nuclear industry. Beginning in 1980 or so, some light-water reactors may be fueled with recycled plutonium. And breeders could produce additional fuel supplies in the form of plutonium as they generate heat for the production of electricity. Bred plutonium could extend the era of nuclear fission devices well beyond the point at which cheaply recoverable uranium is depleted.

Assuming that political, economic, and bureaucratic commitments to the costly breeder program do not divert funding from fusion research between now and the year 2000, it seems likely that fusion plants could overtake fission-based reactors as the basic source of nuclear electric power during the second quarter of the twenty-first century. Tokamak-type fusion devices promise to become the dominant form of fusion technology. However, fusion research proceeds on several fronts. So controlled thermonuclear power may be produced by laser machines (see Figure 4-3) or any of several other now experimental reactor forms.

and further into the future. But the preponderance of expert opinion—represented by the figures in Chapter 2—suggests that the costs of coal-based electric power will continue to exceed that of nuclear power through the 1980s.

Finally—and perhaps decisively against strategies for accelerated burning of coal—there is the fact that this fuel has a more valued use as a feedstock to synthetics plants than as a boiler fuel. Nuclear power cannot be used to drive internal combustion engines. But coal can be so used, provided that it is converted into synthetic fuel. Nuclear plants can provide the energy needed to drive such conversion processes. In other words, coal is far better regarded as a *complement* to nuclear energy than as a substitute for it.

Solar energy probably offers the preferred long-term alternative to nuclear fission, and perhaps to fusion as well. But solar energy remains experimental. In any case, the cost projections for solar power are not particularly encouraging.

Solar power can, it is true, usefully supplement the traditional kinds of home heating and cooling systems. But even if half the houses in America derived fully half their heat from solar energy, the sun would still supply no more than 2 percent of the nation's total energy demand. Even more telling against solar power, the breakthroughs are not in sight that would permit reliance on the sun's rays to provide concentrated central electric generation. Yet it is central electric generation—at which nuclear units excel—that is needed to meet the needs of industry and mass transit. The best foreseeable central solar electric systems may run to five times the installation costs of fission and fusion plants.[1]

Work remains urgently needed to develop the potential of solar technologies. But meanwhile, nuclear power must remain an essential element in America's energy supply strategy. Even if energy growth were held to zero in per capita terms, population growth portends a rising demand for power.[2] In 1975 there were some six million more Americans in the prime childbearing cohort (ages 20 to 29) than only five years earlier. By 1985 the number in the same age bracket will have jumped another five million, a consequence of the boom in births of the late fifties and early sixties. Even with a birth rate of two children per family starting in 1970, substantial population growth seems inevitable. Some 50 to 70 million Americans, demographers predict, will swell the population by the year 2000.

The fact is, then, that America needs more energy, especially in the form of electricity. Nuclear power offers the surest, cheapest, and safest way known to supply it.

If nuclear power were not needed on grounds of sufficiency of supply, it would still be prudent for the United States to maintain a dominant influence in the formulation of international energy policy. The United States can shut down its own nuclear industry, but it cannot put the atomic genie back in the bottle. In 1975 America's share of the world atomic commitment, expressed in the capacity of plants in service or on order, amounted to only 60 percent of the total nuclear-electric outlook.[3] (For comparison, this total commitment equaled the capacity of *all* the hydroelectric power developed to date around the world.) France and the Soviet Union have working breeder reactors. Plutonium reprocessing plants are commissioned or slated in India, Japan, and at several points in Europe.[4]

It is true that by abandoning America's nuclear enterprise, decisionmakers could offer some extra protection to citizens of the United States in the form of an extended buffer zone against the effects of reactor accidents. All atomic plants would then be abroad. "Abroad," however, could be no farther away than Canada, where the commitment to nuclear power seems irreversible.[5] Moreover, the complexity of the world's ecocycles, combined with the persistent nature of the nastiest radioactive materials, means that the problem of nuclear safety cannot be solved merely by halting the advance of the industry in one country. Are we to fear nuclear-materials thieves? A would-be plutonium terrorist could just as readily–*more* readily–secure bomb-grade materials from a Spanish or Egyptian plant as from an American plant. Are we to fear radioactive wastes? Long-term pollution of the ecocycles would as easily result from lapses by Russian or Korean nuclear managers as by American managers.

In 1974 India used plutonium from an imported Canadian reactor to make a bomb. In 1975 the West Germans sold a reactor system to Brazil, but with unprecedentedly relaxed safety and safeguard provisions. The Indian bomb project or the Brazilian purchase must not set a pattern. Yet by opting out of the world's nuclear future, the United States could only relinquish any chance of influencing international safety practices. The United States, by maintaining a strong nuclear program and carefully regulating the American-based corporations to which other nations naturally turn as suppliers of nuclear technology, must take a lead in the development of international atomic controls.[6]

On the grounds of both supply sufficiency and international prudence, then, it can be asserted with some sureness: There *will be* nuclear electricity–more and more of it–in your future. But the prospect of continued development of the atom responds to the problems of energy supply only by exacerbating another problem, the problem of a bitter and erosive dispute between antinuclear and pronuclear spokesmen. A nuclear policy as big as the issues it would address must reckon with the real and serious objections of opponents, even if it cannot satisfy them in terms of ultimate policy recommendations.

What are the facts of the nuclear form? What are the uncertainties? What are the factors that should be appraised in weighing the risks? The answers to these questions, which are developed in Part I, must decisively influence the institutional and socioeconomic aspects of a national nuclear policy, which are addressed in Part II.

Notes

1. Hans Bethe, "The Necessity of Fission Power," *Scientific American* (January 1976), 21, pp. 23-24. See for somewhat more optimistic estimates,

Jerry Grey, ed., American Institute of Aeronautics and Astronautics, *Solar Energy For Earth: The AIAA Assessment* (New York: AIEE, Draft Final Report, April 21, 1975), especially Chaps. 1-2, 5-8.

2. The Report of the Commission on Populaton Growth and the American Future, *Population and the American Future* (Washington, D.C.: U.S.G.P.O., March 27, 1972), pp. 16-24.

3. Joel Primack, "Nuclear Reactor Safety: An Introduction to the Issues," 31 *Bulletin of the Atomic Scientists* (September 1975), 15, p. 17. Further data on nuclear commitments in the United States and abroad will be found in *Info* (New York: Atomic Industrial Forum), releases for May 19, 1975 and November 30, 1975.

4. An excellent survey of U.S. and foreign breeder programs appears as Lawrence Minnick and Mike Murphy, "The Breeder: When and Why," 1 *EPRI Journal* (March 1976), 6. Further information on foreign breeder reactor development is available in John H. Douglas, "The Great Nuclear Power Debate (2); Breeder Reactors," 109 *Science News* (January 24, 1976), 59.

5. See A.J. Mooradian and J.A.L. Robertson, "CANDU Reactor Strategy: Toward a North American Strategy," Chalk River Laboratories, Ontario: Paper delivered at the Cornell-Carleton Conference on North American Energy Policy, October 2-4, 1975, Carleton University, Ottawa.

6. See generally, Mason Willrich, *The Global Politics of Nuclear Power* (New York: Praeger, 1971).

1 Nuclear Technology: The First Two Generations

On December 2, 1942 in a squash court beneath Stagg Field Stadium at the University of Chicago, a team of nuclear physicists under the leadership of Enrico Fermi demonstrated the feasibility of a controlled nuclear reaction. Fermi used a carefully designed pile of graphite interlaced with a critical mass of natural uranium. The pile generated little heat and no light—although it might, if it could somehow have been connected to an electric circuit, have lit perhaps a 100-watt bulb.

A pile such as Fermi's maximizes the likelihood that a neutron emitted from a nucleus of fissionable uranium in the course of radioactive decay will collide with a second fissionable nucleus. As the bombarded nucleus splits, releasing further neutrons and heat, some of the released neutrons should collide with still other nuclei, producing a chain reaction.

Whether a given neutron will in fact thus collide with a fissionable nucleus depends on its speed and on the characteristics of the pile. Neutrons must be slowed (to what are known as "thermal speeds") by a moderator or else they will simply fly from the region of the reaction. In Fermi's pile, the graphite served this moderating function. Moreover, if too few atoms of a fissionable element are packed into the reactor core, a neutron, even if properly slowed, may either fly through the available nuclei without colliding with any one of them or else be absorbed by some nonfissile material in the reactor core (e.g., by the carbon in the graphite of a Fermi-type pile). The critical mass needed to sustain the chain reaction requires the aggregation of enough fissionable atoms in a single location to assure that nuclei will continue to be split by neutrons released in prior fissions.[1]

Most American civilian reactors are moderated by "light water." Light water also carries any heat that cannot be converted into electricity away from the region of the reaction. (Light water or normal H_2O is contrasted with the "heavy water" of Canadian reactors. To make heavy water, oxygen is combined with deuterium, an isotope of hydrogen that contains one extra proton. Hence heavy water or D_2O weighs slightly more per molecule than does ordinary water.)

The Westinghouse Corporation's pressurized water reactor (PWR) burns zirconium-clad fuel elements. In these fuel rods, the most readily fissionable uranium isotope, uranium-235 (U-235), has been enriched to about 3 percent of the total fuel content, up from the 0.72 percent found in natural uranium. Fissioning within the core releases heat which raises the temperature of water circulating in a closed loop under about 2000 pounds per square inch of pressure. Water in the second loop of Figure 1-1 rises to steam as it picks up the heat from the first. The steam expands through turbines, which spin huge magnets through coiled wire to generate electricity.

In mid-1975, 30 commercial PWRs were operating or licensed in the United States, with another 121 units under construction, ordered, or planned.

In the General Electric Corporation's boiling water reactor (BWR)–23 of which were operating or licensed in 1975, with another 53 ordered, planned, or under construction–pressure within the first loop of circulating water is about half that in the Westinghouse design. The lower internal pressure permits the water, as it acquires heat from the reactor core, to boil into steam which then drives the generating turbines. In effect, the first and second loops of Figure 1-1 are combined in the BWR. The boiling water reactor is less complicated than the PWR. But it is also a full percentage point or so less efficient. And the BWR has the further disadvantage of using irradiated steam directly from the core to generate electricity.

In both the PWR and the BWR, the expanded steam must be condensed for recycling. During condensation it gives off the "heat of evaporation" that was used (either in the heat exchanger of the PWR or in the reactor core of the BWR) to raise water to steam at the start of the cycle. This heat of evaporation, plus any energy that was not converted into mechanical motion of the turbines, must be borne off from the plant. In most nuclear plants, at least 2 megawatts are wasted for every one translated by the turbines and generators into electricity. Hence reactors may be rated either in terms of overall *thermal* or total usable *electrical* output, respectively as Mwt (megawatts thermal) or Mwe (megawatts electrical), with the latter roughly translatable into the former by multiplying the usable electric output by three.

In the condenser of a nuclear plant cooling water receives any waste heat and bears it away to a nearby natural water body, to an artificial cooling pond or channel, or to a cooling tower.[2] Ultimately, this waste heat will be transferred to the local atmosphere. Evaporation of water is thought to account for 50 percent of all thermal transfer to the atmosphere, with about 1000 Btus of heat being carried off for every pound of water thus "consumed."

Every day the electric utilities withdraw more than 10 percent of the flowing fresh water in the United States for cooling purposes. Power plants represent the most significant point-sources of waste heat, and by far the gravest potential thermal pollution threat. Moreover, two factors portend a substantial worsening of the waste heat disposal problem: the projected growth in centrally generated power by all means, and the expected relative increase in nuclear generation.

Heat dissipation requirements are higher for light-water reactors of current design than they are for fossil-fueled plants by about 1000 Btus per kilowatt-hour generated. The maximum efficiency attainable by any heat machine

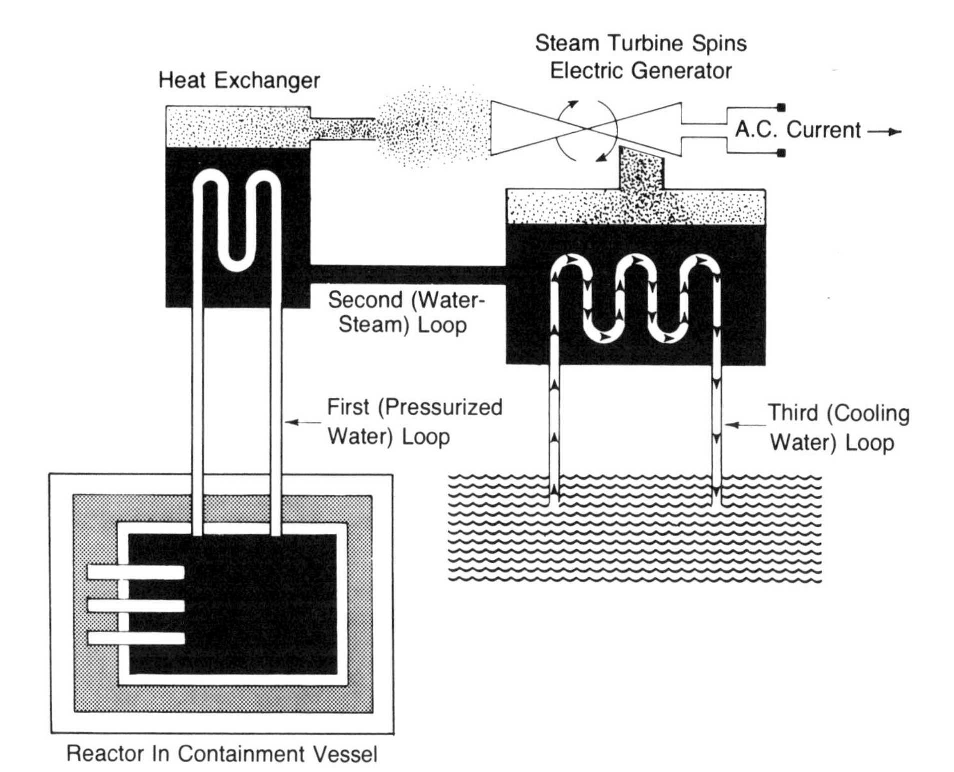

Figure 1-1. Electricity from the Pressurized-Water Reactor.

Notes to Figure 1-1. The Westinghouse Corporation's pressurized-water reactor requires three fluid loops–two closed loops and one open loop. The first loop contains water under pressure to prevent evaporation. This water carries heat from the reactor core to a heat exchanger. There, water in the second core to a heat exchanger. There, water in the second loop picks up the heat and is vaporized. The steam thus generated expands through a turbine that spins an electric generator to produce electricity. Use of a second loop means that the irradiated water from the reactor core never leaves its closed cycle.

The steam is then condensed back into water and pumped once again through the heat exchanger to close the second loop.

The third loop in a PWR is "open" in the sense that it couples the plant to the external environment. In the third loop, a pump pulls cooling water from a source outside the plant and then returns it with the heat of condensation from the spent steam in the second loop to the external source. The cooling water must also carry off any heat that has not been extracted from the steam in the turbine and converted into mechanical energy or friction.

The General Electric Corporation's boiling-water reactor works on the same basic principles, except that the first two loops are in effect combined. That is, water is converted directly into steam by heat from the reactor core. So unlike the PWR, the BWR has irradiated steam in its turbines. This steam is then condensed after recovery from the turbine and pumped back through the core to pick up heat again.

depends on the difference between its highest internal operating temperature and the temperature at which the engine is exhausted to the atmosphere. The higher the internal engine temperature, the higher will be the thermal efficiency of the engine. But with higher temperatures, internal pressures also increase. And although reactor manufacturers go to enormous lengths to assure the physical integrity of all components, high pressures within a reactor can contribute to the possibility of pipe or vessel rupture. To reduce the danger of failure, therefore, light-water reactors are designed to operate at lower pressures than would be optimal from the standpoint of thermal engineering.[3]

Steam in water-moderated nuclear plants will typically be raised to only about 600° Fahrenheit at 1000 to 2000 pounds per square inch, compared with 1050° F. and 3000 to 4000 pounds in coal-fired plants.

In the first generation nuclear plants, then, a premium in the form of significantly reduced plant efficiency is paid to gain an additional increment of operating safety. Since more heat is wasted with the lower pressures of light-water reactors, some thermal pollution from nuclear plants should be seen as a cost purposely sustained to reduce the risk of reactor failure under too-high pressures, with consequent discharges to the environment.

Toward the Plutonium Economy?

A perhaps more severe threat to the viability of the light-water reactor comes from concerns over the availability of ample cheap fuel than from fears of an excessive thermal load on the environment.

The advantage of nuclear over fossil-fired electric power, at least in the eyes of its early proponents, lay in the promise of seemingly inexhaustible energy. A single pound of U-235 has the ultimate thermal potential of almost 1500 tons of coal or 6000 barrels of oil.

But ores rich in natural uranium are in relatively short supply. And with current enrichment techniques (the gaseous diffusion process), it is extremely expensive to concentrate the fissionable isotope U-235—increasingly so, as an expanding nuclear economy pushes miners to extract uranium from lower and lower grade ores.

Basheer Ahmed, an authority on nuclear costs, has estimated that even modest growth in nuclear generating capacity—say, a 4 to 5 percent annual increase extending into the twenty-first century—will produce a yearly requirement for some 35 million tons of uranium oxide fuel by the year 2040. Deposits capable of supporting this requirement are physically present in the crust. But the wastes that would be left after processing the ores might bury much of civilization.

The so-called Tennessee shale grade ores, with only 60 parts per million of U-235, contain enough fissile matter to produce more energy in a reactor than is needed to mine them. (Actually, these ores contain less fissionable material per unit volume than do the tailings discarded from uranium mines in the 1970s.) However, the same amount of electricity could be produced with stripmined bituminous from Appalachia and lignite from western fields, while moving half the total volume of coal as would be needed of uranium ore. What is more, even though reliance on ever-lower grade ores seems feasible in terms of the output from such ores relative to the energy needed to mine them, a move to Tennessee shales would still require an enormous investment of energy, reckoned in absolute terms. Hence, continued production of U-235 to fuel a second generation of light-water reactors may assume the availability of energy that could only be supplied by the very electric plants whose purpose the mining of low-grade ores would be to fuel!

Ahmed's estimates suggest that the nuclear industry will reach a break-even point early in the twenty-first century.[4] At break-even, uranium becomes economically less than competitive with alternative fuel sources. These alternatives could include oil from shale deposits in Colorado, Utah, and Wyoming. And long before oil shales become serious competitors, coal would become a more attractive alternative fuel, even after premiums are paid for coal "benefication" to remove impurities and for environmental controls to prevent any remaining pollutants from entering the ecocycles.

Economic and environmental considerations, then, combine to furnish incentives for developing an alternative to the uranium base. Two interrelated responses—plutonium recycling and the breeder reactor—have been devised.

Plutonium Recycling

As U-235 burns in a reactor, some neutrons released during the chain reaction are captured by the isotope U-238 in the fuel elements. The U-238 is not itself fissionable—not, at least, when bombarded by neutrons that have been moderated to thermal speeds. However, it can be converted through neutron bombardment to uranium-239, which decays (with a 27-minute half-life) to neptunium-239. With a 2.3-day half-life, the neptunium then decays to the fissionable isotope of plutonium, Pu-239. Thus the beginning and the end conditions of this breeding chain appear in the formula:

$$\text{U-238} + \mathit{1\ neutron} = \text{Pu-239}$$

The U-238, the most common isotope of natural uranium, is said to be fertile. Another fertile element, thorium, converts to a fissionable form of uranium, U-233, according to a similar shorthand formula:

$$\text{Th-232} + \mathit{1\ neutron} = \text{U-233}$$

Of course, in the process of fuel burn-up within a reactor core, dozens of fission products and bred isotopes in addition to Pu-239 or U-233 are created. Most of these are not only not useful as potential fuel, but are actually harmful. Unless the core is regularly purged and refueled, the buildup of waste fission products will poison the chain reaction. Moreover, the fuel rods lose their metallurgical integrity as they undergo radiation damage through neutron bombardment. So it is necessary periodically to shut down a reactor, removing spent fuel rods for disposal or reprocessing, in which the elements may be chemically separated from one another.

Since even enriched uranium consists mostly of U-238, the chain reaction will, as long as it continues, breed some plutonium, which will be available in the reactor core. In a reprocessing plant, this fissionable plutonium may be extracted from the reactor wastes to be recycled as fuel back into reactors. (Reactors that breed U-233 from Th-232 are said to be operating on the thorium cycle. Fissile U-233 may be extracted from reactors on this cycle much as Pu-239 may be from reactors on the uranium-plutonium cycle.)

In the preferred mode of plutonium recycling, Pu-239 (extracted from spent fuel elements at a reprocessing plant) and U-235 (enriched from mined natural uranium either by gaseous diffusion or, perhaps eventually, by laser separation) would be combined in fuel rods for loading into light-water reactors.

Plutonium recycling, which gained tentative approval by the U.S. Nuclear Regulatory Commission in early 1975, would reduce by approximately 10 percent the nation's requirements for U-235 as a reactor fuel and thereby extend the usability of the uranium base. But recycling would add enormously to the hazards of nuclear power.[5]

Plutonium compares with botulin and anthrax in the toxicity of even minute doses. Moreover, since plutonium has a half-life in excess of 24,000 years, it remains lethal for longer periods than define the useful lives of any known storage vessels. Recycling could complicate the nuclear safety picture by requiring the transport of plutonium-rich mixed oxide fuel rods from reprocessing sites to reactors for burn-up. Opponents of plutonium recycling fear the physical opening up of the nuclear industry, since a growing transportation network multiplies the sites of potential accidents. It also intensifies the risks of sabotage or diversion. Plutonium, the stuff of the Nagasaki atom bomb, could prove an attractive target of theft for would-be terrorists, urban guerrillas, or militarists in nonnuclear countries seeking the status of a bomb-possessing power.

Is it better, then, to search for ways to store this lethal element virtually in perpetuity, or to burn it up as fuel in other fission reactors? Proponents of plutonium recycling stress that the recycling and, one hopes, the safe use of plutonium from light-water reactors would aid in the development of management techniques and standards applicable to the enormous quantities of plutonium envisioned in an economy based on the breeder reactor.

The Breeder Reactor

In the breeder reactor, the creation of fissionable material as a by-product of core burn-up becomes a conscious element of reactor design. As shown in Figure 1-2, the core is enclosed in a blanket of fertile U-238, arranged to maximize the capture of neutrons. Through nuclear transmutation, neutron capture in the blanket breeds plutonium. Thus in addition to heat, the chain reaction produces new fuel—indeed, more fissionable material than the reactor consumes.[6]

The breeder reactor of primary interest within the United States could make maximum use of U-238, readily available from the wastes of enrichment plants. Hundreds of thousands of tons of these wastes will have accumulated in the United States by 1990. As a source of blanket material, they could contribute an economical source of fuel for centuries. This figure should be compared with the mere 10-percent extension of U-235 supplies possible with the recycling of plutonium from light-water reactors' spent fuel rods.

The breeder reactor calls for significant design improvements over the light-water reactor. Breeding requires a neutron-rich environment, both to sustain the chain reaction and to convert U-238 from the blanket into fissionable plutonium. Even though some plutonium is created in the fuel rods of a BWR or a PWR, the light-water moderator in the reactor core makes it virtually impossible to achieve a full breeding effect. Water not only effectively slows but also actually absorbs free neutrons. Hence the water-steam cycle is an unsuitable one for the advancement of the breeder concept. To overcome the problem of neutron capture, most breeder reactors use a liquid-metal core coolant—hence LMFBR, for Liquid Metal Fast Breeder Reactor.

Sodium, the preferred breeder coolant, has excellent heat transfer characteristics, as well as a relatively low neutron absorption capacity. Moreover, liquid sodium can be used at relatively low internal vessel pressures, even when it has been heated to extremely high temperatures. The pressure needed to circulate tens of thousands of cubic meters of sodium per second through the loops of Figure 1-2, though considerable, falls far below the level of internal vessel pressure that would result if the sodium in a breeder underwent the phase changes (i.e., water-steam-water) of the moderator in a BWR or PWR. For this reason, the temperature of the liquid sodium can be raised without jeopardizing the integrity of the reactor vessel.

Because higher internal operating temperatures mean higher efficiencies, the LMFBR not only promises to solve any foreseeable fuel shortfall associated with the depletion of easily recovered U-235 but also promises to reduce the threat of thermal pollution associated with the light-water reactor.

On the other hand, the benefits of the breeder reactor do not offer themselves as altogether unmixed blessings. Any breaching of the LMFBR core-containment system—*if* it should occur—could release, in irradiated form, one of the most highly reactive substances known. Vented sodium would

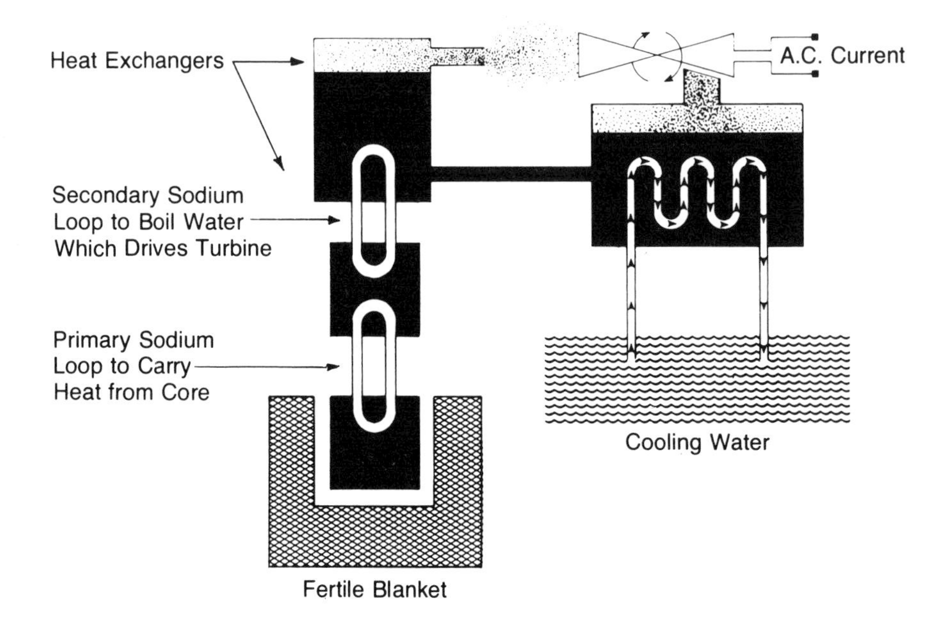

Figure 1-2. The Liquid-Metal Fast Breeder Reactor (LMFBR).

Notes to Figure 1-2. The generating end of an LMFBR–the end which actually produces the electricity, as shown on the right–is identical to that of a light-water reactor, except that flowing sodium rather than water delivers heat from the reactor core to raise steam for expansion through the plant's turbines.

However, on the heating end to the left, there are two main differences between a breeder and an LWR.

First, a blanket of fertile uranium or thorium is carefully arranged about the reactor to absorb neutrons that are thrown off by the chain reaction. These neutrons will transmute the blanket material into fissionable reactor fuel (plutonium-239 or uranium-233).

Second, instead of using water both to moderate and to cool, American LMFBRs of current design use two loops of liquid sodium. Sodium does not undergo the phase changes and consequent internal pressure increases of the water-steam-water cycle. Therefore the breeder should operate at higher exhaust temperatures and with efficiencies about equal to those of the best new fossil fuel plants.

But liquid sodium is a highly reactive chemical substance. Any venting to the environment, due to a flaw in circulation tubes or in the breeder's containment vessel, could have an explosive effect, inhibiting repair and increasing the uncontrolled dispersion of highly irradiated wastes.

Some nuclear engineers–including the influential former Director of Oak Ridge National Laboratory, Alvin Weinberg–would prefer to develop the so-called MS (molten salt) breeder design. In the MS reactor, the fertile blanket is dissolved in a hot fluid, which also serves as both moderator and coolant. The MS breeder has the advantage of design simplicity. But it would tolerate malfunctions less well than the LMFBR. Because reliability figures with increasing importance in American nuclear planning, the U.S. breeder program has concentrated almost exclusively on the liquid-sodium design.

explode immediately on contact with air—spreading radioactivity and enormously complicating the process of clean-up.

As the late 1960s saw the development of "bullish" predictions for the future of nuclear power, the breeder reactor became the highest priority American energy program. Despite the challenge of complex design, and despite the possibility that the breeder would represent a serious environmental or public health threat, it was widely assumed that by 1990 or so the LMFBR would have made a significant impact on the commercial nuclear industry. Unfortunately, experience with the demonstration Clinch River breeder reactor, planned for construction near Oak Ridge, Tennessee, suggests that the early optimism may have been misplaced. Technical difficulties delayed development. Projected costs tripled for the Clinch River plant. The attractiveness of the LMFBR as a solution to the threat of a uranium shortfall diminished. Members of a Harvard-M.I.T. energy study team showed that continued mining of even high-cost (because low-grade) uranium ores for light-water reactors might represent a preferable strategy.[7] With time, the economics of the trade-off between breeder capital costs and low-grade U-235 mining for light-water reactors seemed to tilt ever more pronouncedly against the LMFBR.

Summary: Toward a Strategy for the Transition

The first generation of American nuclear electric stations—based on light-water reactors produced by two dominant manufacturers—seems fated to reach the limit of its possibilities within a half-century or so of its inception. Owing principally to the increasingly adverse economics of uranium recovery, the first generation will probably last no longer than a few decades—only a fraction of the time span during which other dominant fuels of the past (e.g., oil and natural gas, not to mention coal or wood) enjoyed their respective periods of primacy in industrial societies.

Early in the history of the industry, engineers and businessmen who foresaw the end of the uranium-based light-water reactor era laid plans for a nuclear stretch-out through plutonium recycling, and then for a second generation in the development of breeder reactors. But the hazards of a plutonium economy—implicit in both recycling and LMFBR-based electric generation—call the desirability of a second nuclear generation into question.

A breeder bypass cannot by itself answer the concerns of nuclear opponents—this because a reversal of current trends in the civilian atomic industry will not necessarily affect the determination of nations (including the United States) to proceed with plutonium production for military purposes. America's military plutonium inventory is considerable—already large enough to require a concerted effort to solve the long-term radioactive waste storage problem, whether or not the United States moves to plutonium recycling for civilian reactors. U.S. Nuclear Regulatory Commission figures indicated an accumulation to 1975 of some 90 million gallons of military high-level wastes containing plutonium, as against 200,000 gallons of such wastes from civilian atomic plants.[8] The relatively modest increments to plutonium stores that would be added by continued development of light-water reactors (LWRs) may be subsumed to the problem of plutonium management, which must in any case be faced as a consequence of past military programs.

However, a full-blown breeder program would add a major new source of plutonium, and one likely in time to produce quantities that will dominate the "radwaste" control effort.

The implications seem clear. A responsible nuclear strategy presupposes more vigorous efforts to scrap the LMFBR in favor of stretched-out reliance on LWRs. It also presupposes an effort to deal with the issues of light-water reactors that have emerged over the years—and they are very real issues indeed. Two problems have emerged to structure the debate over America's commitment to the light-water reactor: first, the problem of skyrocketing nuclear costs; second, the hazards of radioactivity. Chapters 2 and 3 deal with these critical issues in turn.

Notes

1. On the general characteristics of nuclear technology, see John C. Fischer, *Energy Crises in Perspective* (New York: John Wiley & Sons, 1974), Chapters 9-14; and Chapter II of the 1970 update, Federal Power Commission, *National Power Survey* (IV, pp. 1-31 through 1-70).

2. A standard reference work on thermal effects is Frank Parker and Peter Krenkel, *Physical and Engineering Aspects of Thermal Pollution* (C.R.C. Press, 1970). Also useful are the pieces in *Scientific American* by John R. Clark, "Thermal Pollution and Aquatic Life" (March 1969) and William P. Lowry, "The Climate of Cities" (August 1967). These papers, with others, have been collected in Paul Ehrlich, John Holdren, and Richard Holm, *Man and the Ecosphere* (San Francisco: Freeman, 1971).

3. See my *Energy, Ecology, Economy* (New York: Norton, 1972), pp. 160-162.

4. "Role of Nuclear Power: 1985-2040," 95 *Public Utilities Fortnightly* (May 22, 1975), 21, pp. 23ff. See also M.C. Day, "Nuclear Energy: A Second Round of Questions," 31 *Bulletin of Atomic Scientists* (December 1975), 52, *passim*.

5. See Gus Speth et al., "The Fateful Step," 30 *Bulletin of the Atomic Scientists* (November 1974), 15, and "Plutonium recycle or civil liberties? We can't have both," 12 *Environmental Action* (December 7, 1974), 10; and David

Burnham's excellent survey in the February 27, 1975 *New York Times.* Robert Gillette's two articles are also useful: "Questions of Health in a New Industry," and "Watching and Waiting for Adverse Effects," 185 *Science* (September 20 and 27, 1974).

6. The article by Glenn Seaborg and Justin Bloom, "Fast Breeder Reactors," *Scientific American* (November 1970) presents the technological case for breeders.

7. Irvin C. Bupp and Jean Claude Derain, "The Breeder Reactor in the U.S.: A New Economic Analysis," 76 *Technology Review* (July-August 1974), 5. See also Manson Benedict, "Electric Power From Nuclear Fission," 27 *Bulletin of the Atomic Scientists* (September 1971), 8, pp. 12-15; and the AEC study, "Cost-Benefit Analysis of the U.S. Breeder Reactor Program," WASH 1126 (April 1969).

8. Figures cited by U.S. Nuclear Regulatory Commissioner Edward Mason. See *U.S. News & World Report* (March 29, 1976), p. 56.

2 Problem Number 1: The Economics of Atomic Energy

Early proponents of nuclear energy looked confidently to the day when electricity from the atom would actually prove too cheap to meter, when power from the second sun would become a veritable "free good." All could share in the conveniences of a nuclear-electric wonderland.

Such previsions of the millennium have been cruelly frustrated in the event. The record reveals a steady upward trend in reactor costs, compounded by a punishing increase in the length and rigor of the regulatory review process. And in energetic terms, it has been charged that the extreme complexity of a nuclear plant equates to a requirement that more energy be invested in reactor manufacture and servicing than can ever be paid back, even by a generating station operating at high efficiency and reliability.

Nuclear Power and the "Capital Crunch"

The problem of rising nuclear costs may be considered as the most serious component of a more general cost escalation afflicting the electric industry as a whole. A report by Carol J. Loomis in the March 1975 *Fortune* predicted a need within the power industry to generate some $650 billion in capital between 1975 and 1990.[1] The burden of capital formation must fall primarily on the private electric companies that produce four out of every five kilowatt-hours consumed in the United States. These investor-owned utilities ("IOU's") must enter the money markets and compete against other claims on available capital–including, of course, demands to fund urban renewal and mass transit projects, health facility expansion, and other programs near the top of the social agenda. The remainder of needed capital comes from retained utility earnings.[2] A utility can retain earnings, however, only if it makes them in the first place. Some financial experts question whether the IOU's can break even on their investments unless state regulatory commissions permit rate increases of up to 25 percent per year.

Municipally owned companies and public corporations, such as TVA, raise most of their capital through the sale of tax-exempt bonds. Sellers of these securities compete for available funds against the IOU's, and at a decided advantage in terms of the return they offer on investment. Private utility spokesmen argue that the "unfair" tax-exempt status of their competitors' paper further worsens the industry's ability to fund needed plant expansion.

Four months after the *Fortune* article, a blue ribbon study group of the industry trade association, the Edison Electric Institute, upped the estimate of capital needs to $750 billion between 1975 and 1990.[3]

These cost figures apply to the electric industry as a whole, but they discriminate most severely against nuclear plants, which require up to a third more initial capital outlay than do fossil-fired facilities of comparable size.[4] Despite nuclear energy's lower fuel and operating costs, some fear that the adverse capital cost comparison will, in a time of high interest rates and stiffening regulatory standards, virtually halt private investment in atomic power.

In the late 1960s, it cost an estimated $140 million to build a single 1000-megawatt light-water plant. By the mid-1970s, direct construction costs had escalated to $250 million or so, not far from twice the total cost estimate of only a half-dozen years earlier. Moreover, between 1965 and 1975 another set of cost add-ons, chargeable mainly to tougher safety and environment-related requirements, had loaded yet another $55 million onto the estimated cost of a reactor. In all, the final estimated cost of that 1000-megawatt reactor, hypothetically designed in the mid-1970s and initially slated for commissioning in 1981, rose to $520 million—almost three times the initial contract-price of a reactor that would have been ordered in 1967 or 1968, and intended for commissioning in 1973.[5] A single 1000-megawatt plant will, by the mid-1980s, carry a price tag in excess of a billion dollars.

Table 2-1 presents a comparative assessment of nuclear and coal-fired electric generating cost levels likely to be reached in the 1980s. The figures represent a composite report, based in part on studies conducted in the early 1970s at several leading research institutions, particularly the Stanford Research Institute, Palo Alto, California. Although these estimates have large margins of error, they suggest the extreme sensitivity of the nuclear form to escalating capital costs.[6] A 50-percent increase in capital costs should add only 27 percent to the total sale price of a kilowatt-hour of coal-fired electricity but would increase the price of nuclear by more than 38 percent. On the other hand, coal-fired power costs depend most sensitively on fuel cycle factors (9.65 mills per kilowatt-hour versus 3.85 for nuclear) and on the environmental costs associated with the mining and combustion of coal (4.94 mills per kilwatt-hour versus about a half-mill for nuclear).

The estimates in Table 2-1 suggest that overall, in terms of average cost per kilowatt-hour consumed, nuclear power shows a comfortable advantage over coal—but an advantage that accrues only over the full life of the plant. Alas, the commitment of capital funds to build that plant must precede any payout, and regulatory agencies allow utilities to collect revenues only to cover the costs of electricity that is actually being produced. A paradoxical situation results. Every decision for a fossil-fired plant not only worsens the environmental impact of energy production but also increases the long-run costs of electricity. Yet every

Table 2-1
Comparative Costs of Fission- and Coal-Fired Electric Plants

	Estimated Mills per Kilowatt-Hour	
	Coal-Fired	*Nuclear Fission*
1. Capital Cost of Plant	18.26	25.1
2. Transmission, Distribution	1.1	1.1
3. Full Cycle Costs		
Fuel Preparation		
Mining, Milling, Enrichment or Benefication	7.2	2.0
Fuel Fabrication	–	.35
Transport to Plant	1.2	–
Fuel Recycle		
Reprocessing w Credit	–	.1
Spent Fuel Transport	–	.15
Inventory Charge (16%)	1.25	1.25
4. Primary Environmental Effects		
Land Degradation	.2	.003
Water Pollution		
Thermal	.4	.5
Other	.04	–
Air Pollution	4.3	–
5. Other External Costs		
Sabotage and Diversion	–	.053
Accidents, Explosions	–	.002
Occupational Hazards	.2	.01
Low Level Radiation	–	.001
6. Transport, Disposal of Wastes		
Radwastes (High Level)	–	1.06
Low Level Wastes	–	.1
	34.15	31.779

such decision eases the toughest immediate problem—that of securing the needed initial capital or "front money."

Licensing and Regulatory Delay

Further contributing to the increase in nuclear power costs is the lengthening process used to license new units. The period between a utility's decision to build and the actual start-up of an atomic station stretched from approximately five years in 1965 to eight, nine, or even ten years in 1975.

The most thorough study of the period from blueprint to start-up of a new nuclear plant, by a Harvard Business School-M.I.T. team under the leadership of Irwin Bupp, has shown that the licensing phase—not actual construction—accounts for the major time delays.[7] Moreover, the Bupp analysis showed that the vast majority of cost increases can be traced more or less directly to these delays—that is, to problems encountered in the gaining of regulatory approval for a proposed plant site and its associated generating unit design.

How are these findings to be explained? An electric utility normally operates as a "natural monopoly." Since within its franchise area the firm does business without economic challenge from competing utilities, rate regulation and certification of new facilities by the appropriate state public service commission prevents the utility from extracting monopoly profits. The requirement to gain permission from a regulatory agency before starting construction on a new electric station reflects the determination that utilities are "affected with a public interest." Since utilities operate largely unchecked by private competition, they must show that new facilities are needed and will be safe before an agency can issue a "certificate of public convenience and necessity."

But since the mid-1960s, the regulatory picture has become far more complicated. The federal Environmental Protection Agency emerged to enforce air and water quality standards. Federal judges began pronouncing new rulings, beginning with the 1962 *Scenic Hudson* case, in which the obligation of utilities to consider aesthetic and environmental impacts of power plant siting became a legally enforceable rule of law.[8]

The extension of the regulatory process at the state level, in the federal bureaucracy, and in the courts established the procedural framework in which the licensing delays of the 1970s became all but inevitable. Given the current confusion of regulatory bodies, a utility seeking permission to break ground for a new generating station may have to deal with upwards of 20 agencies of government. Even after the site has been approved and then plant construction completed, a further review process must precede the start-up of operations. The regulatory gamut at both the federal and state levels must be run again and again—each time a utility wants a new unit, or at least each time it seeks to develop a new site as a generating station.

Some modification of regulatory processes seems in order whether or not the electric industry moves toward concentrating reactors in nuclear centers. Current review procedures, which are cumbersome enough in themselves, actually invite delaying tactics by those who oppose development. The existing procedures have innumerable points at which an opponent, by intervening in a licensing proceeding, may bring the administrative process to a temporary halt. Any shortening of the regulatory review process would tend to reduce the costs of electric power, if only by cutting legal fees, by reducing the time spent in hearing rooms during government proceedings, and by shortening the period from blueprint to construction during which inflationary cost increases can accumulate.

Efforts have been undertaken in several states (with New York, Maryland, and Washington leading the movement) to simplify licensing procedures by providing "one-stop" regulatory service. Instead of running a complex regulatory gauntlet, a utility may gain all needed approvals for plant construction by meeting the requirements of a single coordinated process.[9] But until one-stop programs exist throughout the nation–and perhaps not until procedures have been developed for rapid review of siting plans by multistate regional bodies–the cumbersome licensing process will continue to cause delays in construction, uncertainties in planning, and worsening inflation in the costs of electric power production.

Licensing can be expedited, however, only when a broad national consensus develops on the value of nuclear power. "The problem here," Bupp has concluded, "is not administrative or technical in the sense that a somehow inefficient set of regulatory machinery has caused 'artificial' increases in the costs of nuclear power."[10] Rather, the numerous opportunities which the American administrative system affords for delay have been used to reflect severe doubts over the value of atomic energy–doubts that are harbored by groups of significant size and influence.

To be sure, the lengthening of the licensing process has directly added to the cost of power in several ways, e.g., higher legal fees and a stretching of construction lead times in a period of general inflation. But the lengthening of this process betokens even more significant indirect impacts–as when delay results from an extensive debate, forced by power plant opponents, leading to a requirement for costly added safety features for the facility under review. Such cases are by no means atypical. The new safety device required as a consequence of more thorough or more demanding review–rather than regulatory delay itself–contributes to the increase in plant costs.

References to the regulatory process as a principal obstacle to the advent of nuclear power must therefore be put in context. Neither the cumbersomeness of litigation nor the mere fact of regulatory lag can be held accountable for the dramatic escalation of nuclear costs. But the term "licensing delays" can fairly be used as a kind of proxy term, referring to the full array of delaying and corrective opportunities that are provided in the American political system to protect minorities whose spokesmen feel threatened by a proposed course of action.

The "capital crunch" derives from causes largely beyond the control of nuclear technologists. But the problem of regulatory delay could prove more amenable to solution, given a broad-scale effort fully to acquaint the American people with the advantages and the disadvantages of the nuclear form. The broadening base of popular opposition to nuclear power, from which the hard core anti's draw support, has been based on suspicions, often fostered by the indirections or half-truths of nuclear proponents themselves, that atomic plants are less safe, less productive, or less environmentally benign than Americans were initially led to believe. Hence regulatory delay has been used to force nuclear engineers and utility executives to do what they should in many instances have done without official prodding–that is, to justify fully their claims of an accelerating need for additions to generating capacity; more adequately to ensure the safety and the environmental compatibility of proposed new units; thoroughly to consider the induced economic effects of a new construction project throughout the locale of the site.

In Bupp's summary words, "Present trends in nuclear reactor costs can be interpreted as the economic result of a fundamental debate on nuclear power within the U.S. community."[11] In this debate, facts and balanced judgments—not extravagant claims for atomic power on one side, or antinuclear ideology on the other—should ultimately prove decisive.

The Energetic Costs of Nuclear Power

The deteriorating cost picture requires some attention to the energetic, as well as to the pecuniary, requirements of nuclear development. A light-water reactor represents one of the most complex and costly of man's high technologies. To build a reactor requires a painstaking fabrication process by highly skilled workers. It requires the acquisition of exotic metals for use in the core or the containment vessel. To mine a metal takes energy; to ship it takes more energy; to shape and alloy it for installation requires yet a further energetic investment.

The equivalent of more than 2 billion kilowatt-hours of thermal energy are needed to make the steel and the concrete that go into a light-water reactor.[12] Another 3 billion kilowatt-hours thermal are needed to mine and enrich the initial fuel loading. Following start-up, the equivalent of some 5 percent of the reactor's electrical output will be needed to prepare replacement fuel elements over the life of the plant. Finally, as much as another 2 billion kilowatt-hours must be provided to build transmission and distribution networks for delivery of electricity from the reactor. In sum, the energetic price tag for 1000 megawatts of nuclear electrical capacity lies in the range of 7 billion kilowatt-hours thermal.

Can the electrical pay-out of that reactor possibly justify such an initial energetic investment? Ideally, the dollar costs of a new nuclear plant would measure the true value of all capital, labor, and land—including the energy which these factors incorporate—used in the project. A reactor should start returning dollars on its investment the very day that it begins producing electricity. Were the present worth of the expected total discounted cash revenues not larger than the initial investment, the plant could not have been financed to begin with. But in fact, the market can be a notoriously imperfect indicator of true values—especially in a society whose entrepreneurs have traditionally undervalued natural resources. Hence computations have been performed to indicate the actual energetic returns of a typical nuclear plant, to weigh against the energetic costs incurred in building and fueling it.

A 1000-megawatt unit operating at 60 percent of capacity (probably a conservative estimate) over a 30-year plant life will produce almost 160 billion kilowatt-hours of electrical power, about seven times the total thermal energy required to bring the plant on line in the first place. So if it were possible to convert all the electricity from a nuclear plant into the thermal forms that are needed when commissioning a reactor system, a nuclear unit over a 30-year life could "pay back" all the energy used to build that unit and additionally support the construction of six more like it.

But a crucial qualification applies to this seven-to-one estimate. The kilowatt-hours expected during the final years of a nuclear plant's useful life cannot be "borrowed" in advance of their production in order to meet needs encountered in the early years. Hence if additional plants are to be built early in the life of the initial plant, energy supplements from other sources will be needed. Since thermal rather than electrical forms of energy are mainly required to build new plants, the supplement would probably come from the burning of coal (in milling and manufacturing of reactor components) and oil (in transportation).

What really lies at issue in America's decision on the rate of nuclear buildup, then, is the trade-off of an acceleration in the consumption of fossil fuels versus present and future reliance on nuclear power. As will be further discussed in Chapter 8, despite the existence of large coal reserves in the United States, it is not a matter of indifference whether America's consumption of fossil stocks is speeded up. If nuclear development goes too slowly, the nation's energy demands must be met by burning fossil reserves that would better serve future generations as feedstocks to the petrochemical industry (plastics, pharmaceuticals, fertilizers, and so forth). But if development proceeds too rapidly, most of the energetic output of existing reactors during the period of the buildup will effectively be used up to build new plants. In this case, the process of nuclear advance, *not* a growing population or an expanding economy, will get the main benefit of the added capacity.

According to a British physicist, John Price of Friends of the Earth, Ltd., a program that doubles nuclear capacity every 2.5 years could actually represent a net energetic loss to society during the period of the buildup. No country plans such an expansionary program.[13] However, the Nixon energy program, "Project Independence," as updated by Gerald Ford in the 1975 presidential energy message, called for roughly six doublings of U.S. nuclear capacity before the year 2000—a doubling every four to five years. The Project Independence schedule would set forth from a 90,000-megawatt base in 1980, building up to 1,250,00 megawatts of installed capacity in the year 2000. (See the upper curve of Figure 2-1.) The immediate energetic payout to society would be enormously reduced in such a program, and during just those years in which the designers of Project Independence hope that nuclear expansion can help ease reliance on imported Arab oil. By Price's calculations, a four- to five-year doubling time implies that the equivalent of half the nuclear electric output must be directly "re-invested" just to build more plants.

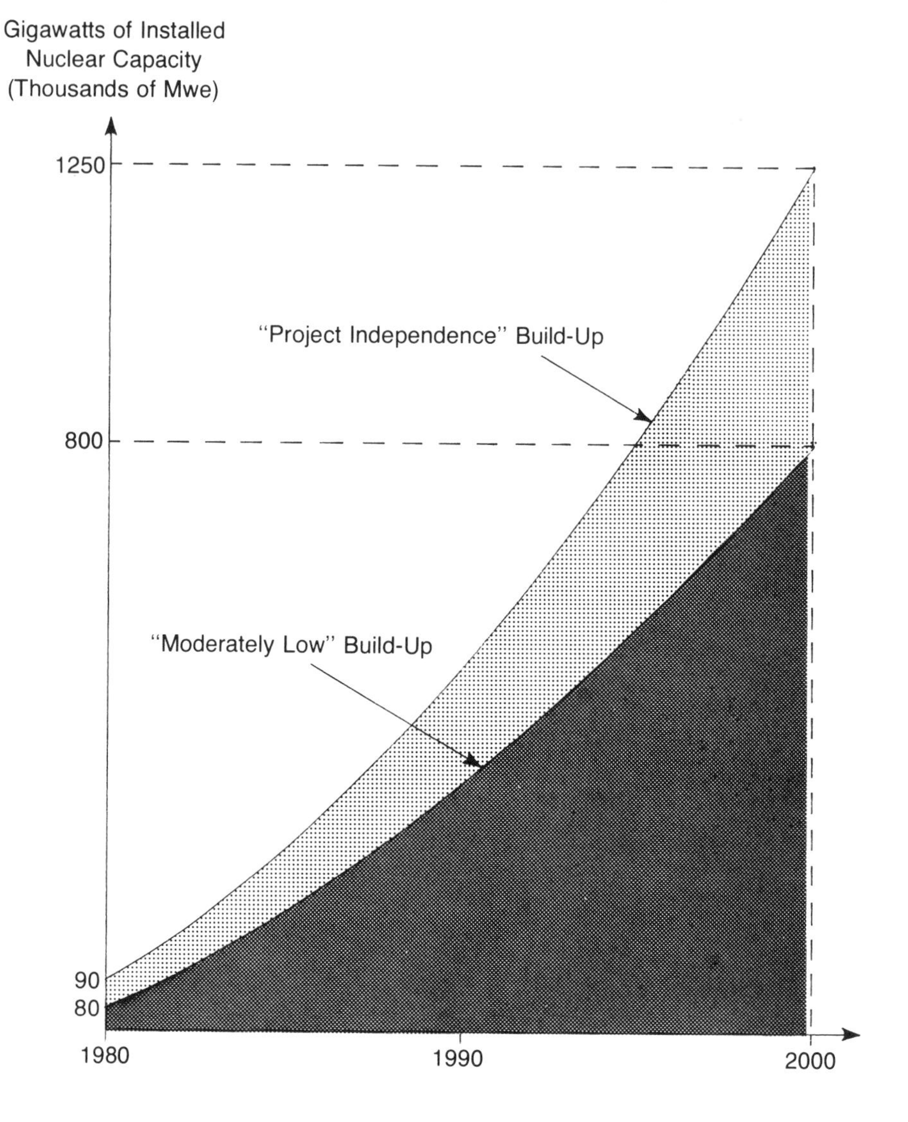

Figure 2-1. Alternative Nuclear Growth Profiles, 1980-2000.

The lower nuclear buildup curve of Figure 2-1 corresponds to a total light-water reactor capacity about 11 percent below the Project Independence target figures in 1980 and 33 percent below the Nixon-Ford targets for the year 2000. These lower estimates are consistent with figures for a "moderately low" rate of nuclear growth as projected by analysts of the U.S. Energy Research and Development Agency in February 1975.[14] Clearly, this reduced rate of buildup represents the preferable course if we desire not to have one more fission reactor in service than necessary when safer electric generating technologies, such as controlled fusion plants, start coming into their own.

Not even the most optimistic estimates of uranium reserves support any belief that 1,250,000 megawatts of installed nuclear capacity can be cheaply fueled with U-235. Hence, most energy analysts associate the Project Independence buildup rate with strong pressures to adopt the breeder reactor. The lower rate of nuclear growth also might create pressures for plutonium breeding. In the worst case—but one accepted as realistic by some students of the problem—uranium reserves might be so low as to require recycling of plutonium when LWR capacity goes much above 400,000 megawatts, and breeding at the 500,000 level or so.[15] But even in these cases, as will be discussed in Chapter 4, a perceived need for plutonium to bridge the gap while waiting for fusion can probably be met with special hybrid systems that link fusion reactors to modified LWRs.

Such a scenario would remain consistent with the preferred strategy for America's nuclear buildup: a strategy to ensure the minimum buildup of fission-based power needed to meet essential energy needs, while investing in the development of more desirable alternatives—fusion and some solar—to begin replacing light-water reactors in the early twenty-first century.

Summary: Toward a Balanced Nuclear Buildup

Needed is a square facing of the fact that nuclear costs, already high, will probably keep going higher. In no quarter is such recognition more badly called for than in the nuclear community itself. Nuclear advocates have consistently undermined their credibility by ignoring or even suppressing the truth about reactor costs.

A new candor and openness among those who would chaperone America into a nuclear future, combined with innovative concepts for dealing with the alarming escalation of atomic plant costs, will offer no formula for inexhaustible, low-cost power. But these departures do suggest the possibility of a national dialogue. From such a dialogue there could emerge a broad consensus on the extent to which nuclear advantages outweigh certain inescapable disadvantages.

A want of such agreement to date, reflected in increasingly cumbersome and unpredictable licensing procedures, accounts for most of those elements of cost escalation that have been charged, erroneously, to inefficiency in the licensing process. Regulatory procedures *do* exacerbate a deteriorating nuclear cost picture—but in return for tightened safety and environmental safeguards, some of which are surely well worth the extra price.

The extra price for such safeguards, combined with escalating costs of money in the capital markets, worked to reduce the number of orders for new plants in accord with standard supply-demand economics. As a consequence, the cost increases in the nuclear field offer little comfort (and small opportunity for profit) to reactor manufacturers who advocate rapid, wholesale reconversion of America's electric industry to a nuclear base.

But the cost increases of the early 1970s may actually encourage those pronuclear figures who see the part of wisdom to lie in a more guarded, gradual buildup of light-water reactors. A reduction in the rate of new plant commissionings may translate into a more balanced nuclear growth program. A reasonable supply of nuclear electrical output can still be made available during the years of transition from fission to fusion—and available, moreover, for uses other than the compounding of nuclear industry growth itself.

Notes

1. "For the Utilities It's a Fight for Survival," *Fortune* (March 1975), 97-189, *passim.*

2. See generally Edward Berlin, Charles J. Cicchetti, and William J. Gillen, *Perspective on Power: A Study of the Regulation and Pricing of Electric Power* (Cambridge: Ballinger, 1974), Chap. 1, in which the technicalities of utility rate-setting are lucidly discussed. See also generally, for perhaps the best introductory discussion of the utilities' financial plight, Mason Willrich, "The Electric Utility and the Energy Crisis, I," 95 *Public Utilities Fortnightly* (January 2, 1975), 22, pp. 24-27.

3. Edison Electric Institute Committee on Economic Growth, Pricing, and Energy Use's *Economic Growth in the Future* (June 1975), available from the E.E.I., New York.

4. A systematic analysis of the impact of rising capital costs (as well as other possible perturbations in the nuclear economic situation) will be found in Paul Joskow and Martin L. Baughman, "The Future of the U.S. Nuclear Energy Industry," under NSF-RANN Grant SI A73-07871 A02), pp. 42ff.

5. Estimates derived from a series of Atomic Energy Commission studies (WASH-1082, -1150, -1235, and further projections) as reported in Figure 9 of a study by United Engineers and Constructors, "Fusion Balance of Plant Information," prepared in October 1974 for the Aerospace Systems Laboratory of Princeton University.

6. "The Economic and Social Costs of Coal and Nuclear Electric Generation: A Framework for Assessment and Illustrative Calculations . . ." based on S.R.I. analyses under contract with the U.S. Office of Science and Technology. See summary in R. Bernardi and B. Barks, *Quantitative Environmental Comparison of Coal and Nuclear Generation, Workshop Summary* (Mclean, Virginia: The Mitre Corporation, September 1975), pp. 17-19, 29. The higher figures shown in

Table 2-1 reflect my own judgments as well as the advice of my student, Robert Brenner. See also B.E. Prince, J.G. Delene and J.P. Peerenboom, "Nuclear Fuel Cycle Economics: 1970-1985," Oak Ridge National Laboratory, October 24, 1975, Division of Reactor Research and Development.

7. Irvin C. Bupp et al., "Trends in Light Water Reactor Capital Costs in the United States: Causes and Consequences," (Center for Policy Alternatives, M.I.T., December 18, 1974); and "The Economics of Nuclear Power," *Technology Review* (February 1975), 15.

8. See my "Environmentalism versus Energy Development: The Constitutional Background of Environmental Administration," 35 *Public Administration Review* (July-August, 1975), 328; and my review of Stephen Breyer and Paul MacAvoy, *Energy Regulation by the Federal Power Commission*, 12 *Harvard Journal of Legislation* (February 1975), p. 293.

9. E. Graf-Webster et al., *Legislation and Regulation of Utility Corridors and Power Plant Siting* (McLean, Virginia: The Mitre Corporation, February 1975 under contract to the U.S. Geological Survey), *passim.*

10. Irvin C. Bupp et al., "Trends in Light Water Reactor Capital Costs . . . ," p. 20.

11. Ibid., p. 21.

12. The quantitative analysis that follows is based on rough estimates in Frank von Hippel et al., "The Net Energy from Nuclear Reactors," *F.A.S. Professional Bulletin* (April 1975), 5. For a different set of figures—one strongly biased against nuclear power—see E.J. Hoffman, "Overall Efficiencies of Nuclear Power" (Natural Resources Research Institute of the University of Wyoming, Circular No. 73, December 1971), at pp. 66ff. On the other hand, more conservative estimates (than von Hippel's) of the input-output energy balances for nuclear reactors are presented in Appendix B, Vol. 1 of E.R.D.A.'s "National Plan for Energy Research, Development and Demonstration, . . . 1976" (ERDA 76-1, April 15, 1976), especially pp. 113ff.

13. My colleague Dr. von Hippel kindly furnished the Price study, "Dynamic Energy Analysis and Nuclear Power" (London: Press-on-Printers, a Friends of the Earth report, December 18, 1974). See especially p. 24.

14. Update of the AEC WASH-1139 (1974) projections of nuclear electric growth, available from the Office of Planning and Analysis, E.R.D.A. The following three studies are also to be particularly commended: F.J. Tyrrell, "Projections of Electricity Demands" (Oak Ridge National Laboratory, November 1973); and Edward A. Hudson and Dale W. Jorgenson, "U.S. Energy Policy and Economic Growth, 1975-2000," for the Ford Foundation Energy Policy Project, *Policies for Energy Equilibrium* (Washington, 1974), 461; Charles E. Whittle et al., "The I.E.A. Energy Simulation Model: A Framework for Long-Range U.S. Energy Analysis" (Oak Ridge Associated Universities, ORAU-125, January 1976), especially Chapter II and Appendix C.

15. Upward revisions of uranium projections in 1975 suggest that the pressure to adopt plutonium recycling may be somewhat less severe. See "The I.E.A. Energy Simulation Model," ibid., p. 148.

3 Problem Number 2: Judging the Risk, Hedging the Bet

Ever since people learned of radiation sickness following the atom bombing of Japan, the threat of long-term harm to whole populations has overshadowed the nuclear enterprise. The probable advance of nuclear technology in the United States, and its certain proliferation throughout the remainder of the world, suggests that *all* people will live in an environment to some extent poisoned by radioactivity. Cumulative somatic or genetic damage can result from persistent exposure to low-level emissions. Nuclear development also raises the possibility of a catastrophic public menace following a major release of radioactivity in a kind of "cloud of death" after a reactor accident.

To what extent can we accurately judge the risk of nuclear power? To what extent can we hedge a bet on the atom by taking special precautions in the design of reactors and the management of radioactive wastes?

In a democracy, responsible choice requires a body politic in command of the facts, when hard facts exist—but aware of uncertainties when factual analysis cannot be relied upon. Alas, the nuclear dialogue has been severely polluted with fraudulent quantitative indices. Efforts to appraise the radioactive exposures of populations under varying nuclear emission standards have yielded some grotesque facsimiles of quantitative estimates. Numbers derived from suspect data fill the literature of nuclear inquiry. For example: One contributor to the nuclear debate alleges that a PWR implies "18,000 times less health risk than a coal-burning power plant."[1] Another example: A nuclear opponent predicts that 1000 light-water plants in America will cause "at least 5,741,500 future deaths from lung cancer over the next 80,000 years."[2] Plainly, contenders on both extremes of the nuclear debate have lost sight of the limits of the knowable.

Quantitative estimates of nuclear hazards can only express rough probabilities. Multiple-digit indices ("18,000 times less risk," no less!) can give only order-of-magnitude comparisons. With such qualifications applied to the data on both sides of the radiation issue, can any general conclusions be drawn regarding the acceptability of fission-based power?

Four categories of potential hazard need consideration: (1) the threat posed by "routine" emissions of low-level radioactivity from nuclear facilities; (2) the danger of radioactive releases as a result of an accident at a nuclear plant; (3) the potential for abuse implicit in diversion of radioactive materials (including both plutonium and high-level wastes) or in their acci ntal release from long-term storage facilities; and (4) the problems posed by bulky low-level radioactive wastes, whose dilute radioactive content invites carelessness in handling and disposal.

Routine Radioactive Emissions

Radioactive wastes from the mining, milling, and processing of uranium ores are routinely left exposed in refuse banks. In addition, certain radioactive gases and liquids are routinely released from operating reactors. Further routine releases occur as spent fuel elements are reprocessed to recover plutonium or recycleable uranium.

That releases are "routine" does not mean that they are desirable. Clearly, it would be preferable to have no radioactive emissions at any level. The word "routine" is a euphemism referring to those (usually minute) quantities of radioactivity that simply would be too expensive, or in some instances impossible with state-of-the-art technology, to keep segregated from the environment.

The unit of measurement for radioactivity is the "rad"–that amount of energy which is deposited in a gram of matter when the equivalent of 100 ergs of radiation hits the material in question. (Because the particles of radioactive decay are minutely small, it may take 100 million impacting particles to produce 100 ergs. By the same sign, in matter which is exposed to low levels of persistent radioactivity, a rad's worth of energy may take years to accumulate.)

The effect of radioactive hits on biological tissue may vary over a considerable range, depending on the kind of particles and the kind of tissue. The energy deposited by an x-ray is used as a unit standard for hits on biological tissue. Technically, x-rays give a measure of "relative biological effectiveness," or RBE. The number of rads carried by a given dose of radioactivity, multiplied by the RBE of the biological tissue being hit, gives the organism's exposure level in "rems."

The average American receives one-tenth of a rem–usually recorded as 100 millirems–per year from natural background radiation (e.g., from cosmic rays that manage to penetrate the earth's protective ionosphere). Medical and dental x-rays typically account for another 70 millirems. The spread of light-water reactors will necessarily add somewhat to the 170-millirem annual radioactive burden from natural and diagnostic sources. A commonly heard figure is one-tenth of an extra millirem for the average man, woman, or child in the United States, given 1000 operating light-water reactors. The addition of reprocessing plants in numbers sufficient to service 1000 reactors might bring this extra general exposure level to the half-millirem mark.[3]

(It hardly needs emphasis that radiation estimates computed on the basis of political boundary lines are somewhat fanciful. Mexicans and Canadians will bear some of the general radiation burden created by American reactors. Conversely, even if the United States were to halt all nuclear development, Americans would bear some of the radiation load attributable to Canadian and perhaps to other nations' reactor programs as well.)

In addition to the general radioactive ambiance–between 150 and 200 millirems per year per person–higher radiation dosages will be sustained by organisms located near the main point sources. These point sources include piles of uranium-bearing mine wastes or mill tailings, as well as working reactors and reprocessing plants.

Mine Wastes and Mill Tailings

Since raw ore contains only a fraction of a percent of the isotope U-235, the preparation of enriched fuel elements supposes a series of separative and concentrative processes. At each stage—uranium mine, mill, fuel enrichment, fuel fabrication—wastes are discarded, leaving behind trace amounts of radioactivity.

Uranium naturally decays by radioactive disintegration (that is, emission of alpha or beta particles) through thorium, pacificum, radium, radon gas, and polonium, until it becomes stable as an isotope of lead. Minute quantities of all these elements, left behind in the wastes of mines and mills, are vulnerable to leeching, because waste piles consist of loosened matrices that can readily be penetrated by rainwater or runoff. Radium that has been washed from a waste pile decays to radon gas. Because it is a gas, radon is mobile. Unlike the radium and polonium which occur before and after it in the decay chain, radon gas needs no additional transport mechanism, such as flowing water, to move it into populated areas.

It is, incidentally, the source of radon in thorium that has given rise to the previously cited estimate of almost 6 million deaths from lung cancer spread over 80,000 years (the effective radioactive half-life of thorium-230). However, such baleful projections—quite apart from the mysteries of the computations that underlie them—make no provision for possible future improvements in cancer prevention and treatment. Nor do they consider the relative ease with which low-level mine wastes and mill tailings can be effectively segregated from the environment long enough to immobilize the fugacious isotope, radon.

Most students of the subject concede that the hazardous contents of mine and mill wastes could be trapped if the tailings were bulldozed over. Radon has a half-life shorter than four days. So if it can be segregated for a relatively brief period, most of the hazard-potential on a given site will thereafter quickly diminish. Landfill entrapment in areas known to be geologically stable and relatively dry might also be applied to the wastes from uranium enrichment plants (unless, of course, such wastes are to be mined for U-238 to be used in the fertile blankets of breeders) and from fuel fabrication facilities.

Wastes from mines, mills, and enrichment and fuel fabrication facilities lend themselves to relatively simple techniques of control because they are characterized by low levels of radioactivity—that is, by trace quantities of dangerous isotopes that are widely dispersed through a benign, nonradioactive matrix. Moreover, mine wastes and mill tailings generally occur in relatively open, and often in remote, regions where land costs are low enough to permit condemnation of large areas for disposal sites. And large areas *are* needed, since with such wastes the volumetric problem dominates. An enormous bulk of materials must be stored or buried in order to ensure segregation of the dangerous elements that are diffused through the matrix.

Nuclear Plant Emissions

Radioactive releases from active reactors and reprocessing plants present more difficult control problems—not because these emissions are inherently more dangerous or more concentrated than are the wastes of mines and mills, but because vented gases or liquids can be less tractable to physical disciplining.

During the burn-up of uranium, the main routinely vented radioactive pollutants are krypton-85 and gaseous tritium to the air and liquid tritium in the plant's outflowing cooling waters.[4] Suitable (though increaingly costly) investments in reactor shielding and containment, as well as in improved waste reprocessing and disposal technology, could reduce free venting of these radioactive substances. In fact, nuclear power has always been characterized by substantial investments in the name of safety. Some of these investments have been made in the knowledge that atomic energy would thereby become less competitive with other fuels, such as coal, whose developers have traditionally shown less concern for the problem of negative externalities. But not even the prodigious returns of nuclear energy can support the investment necessary to give perfect control, perfect safety, or perfect insurance against all future contingencies.

Radioactive Krypton. One hundred and fifty thousand megawatts of Free World fission capacity in 1980—a level roughly consistent with the lower projections of nuclear growth in Figure 2-1—would, with continued uncontrolled venting of krypton-85, imply an additional three-tenths of a millirem exposure per year per person, assuming complete dilution throughout the atmosphere. Krypton-85 has a half-life of almost 11 years, long enough to permit a gradual buildup over time. Because krypton *does* tend toward more or less uniform dispersion, the incidence of cancers traded to this effluent will be spread literally around the world—a point that underscores the fact that a mere American nuclear moratorium could hardly insure citizens of the United States against radioactive hazards so long as foreign nations continue to press their own vigorous programs of atomic development.

Public health authorities, using the higher nuclear growth rate projections that AEC analysts were publishing in the late 1960s (e.g., predictions of 500,000 or so megawatts of Free World nuclear capacity in 1980), estimated a krypton-85-associated dose of 2 millirems per person in the year 2000, and up to 50 in 2060. Even 50 millirems would represent less than 1/500 of the minimum whole body dose needed to produce observable clinical effects in humans.

Nevertheless, most geneticists suspect that radioactive bombardment at *any* level can lead to mutations.

Because the global krypton-85 dose represents but one component of the total radioactive load, the case is strengthened for moving forward with light-water reactors only as part of a more comprehensive program that looks to a downward inflection of the fission-based nuclear growth rate at some definite, foreseeable point. A plan whereby fusion would begin replacing fission in the early twenty-first century represents such a program. As we shall see in Chapter 4, controlled thermonuclear power is not without radiological problems of its own. But these problems are, at their worst, vastly less threatening than are those of fission power.

Radioactive Tritium. Most of the factors that bear on the hazard of public exposure to radioactive krypton also apply to routine releases of tritium. Tritium, with a half-life of about 12 years, does not concentrate in biological tissues. Hence the critical index of potential harm is the so-called biological half-life of tritium, from 8 to 12 days. That is, half a given quantity of ingested tritium will have been eliminated from the body (mostly through the excretion of water) in about 10 days, three-fourths in another 10 days, and so forth.

Atomic Energy Commission studies in the late 1960s, when projections of nuclear development were at their highest, suggested that continued free venting of tritium from reactors and reprocessing facilities would add about one-hundredth of a millirem per year to the average exposure in 1980. This estimate assumed perfect dilution throughout a 10-kilometer atmospheric envelope, in the circulating waters of the hydrosphere, and in the seas to a depth of 40 meters. But, of course, dilution is *not* perfect, uniform, or global. Tritium tends to concentrate geographically. Thus, although local dilution in the air or waters near a nuclear plant can reduce the concentration of tritium, and although, furthermore, airborne or hydraulic transport will usually remove some effluents from the vicinity of release, persons in the immediate locale of a reactor or a reprocessing plant inevitably experience some additional radioactive exposure.

The traditional defense against excessive local exposure entails the dedication of an exclusion area around the facility to limit access by members of the general public to the source of a radioactive release. In addition, plants should be sited with attention both to prevailing wind patterns and to the availability of water for dilution. (Consolidated Edison's three Indian Point nuclear units upwind of New York City stand as horrible reminders of the fallibility of systems planners. Indian Point represents a siting error that need never be repeated.) The taking of sensible precautions when siting future plants, together with careful design of containment systems, can ensure that maximum exposure of an organism constantly at the fenceline around a nuclear plant will never exceed 15 millirems—and should in most cases lie below 5 millirems—per year from routine releases.

Analysts of the U.S. Nuclear Regulatory Commission found that the 5-millirem exposure level can also be achieved when reactors are clustered on a common site to form a power park—and even with a reprocessing plant, the most prolific source of routine radioactive emissions, in the immediate neighborhood.[5]

With fission, maintenance of the radioactive threat near the 5-millirem level associated with routine releases may prove to be the best possible case[6]—something like a lower boundary on the radioactive release level of light-water plants. Accidents, mismanagement of radioactive inventories, diversion of plutonium wastes, inadequate enforcement of physical exclusionary rules that are designed to limit worker exposure—these or other factors could only work to heighten the radioactive threat.

Operation of every nuclear plant at the 5-millirem level or better can perhaps be sustained—albeit with "fear and trembling" lest some lapse should result in a serious reactor accident—for a few years, even decades. During these, the decades of transition from fission to fusion, the most extraordinary engineering and managerial efforts may succeed in keeping all the complex safety systems of a light-water plant functioning exactly to specification.

But as reactors and reprocessing plants multiply, their operation will tend to take on the sense of the routine. And as the term "routine" comes increasingly to describe a psychological outlook as well as an economic-technical euphemism, the likelihood of slackened attention to the details of nuclear safety will increase proportionately. It is then, when the guard goes down, that the price—a reactor breakdown, a major accidental release of radioactivity—could be exacted for the human theft of the fires of Prometheus.

So the central point emerges: A willingness temporarily to suffer a gradual buildup of radioactivity in the global environment as a result of so-called routine releases from nuclear facilities should be premised on a "maximum effort" by the nuclear industry for strictest design, construction, and operating safety standards. Since such an effort cannot credibly be maintained for an indefinite period, further fission development should be conditioned on a firm commitment to a shift in the direction of safer energy forms—solar and fusion—as they become available. This kind of commitment presumes major investments in research and planning to bring these forms into being, and hence to make the shiftover possible.

Solar energy represents a nonradioactive power source, and an extremely promising one for decentralized applications, such as home heating. But as emphasized in the introduction to Part I, major breakthroughs are needed to bring the costs of central solar electric systems into line with the likely costs of comparably sized controlled thermonuclear fusion devices.

Although fusion plants will probably routinely emit radioactivity in the form of minute tritium releases, engineers feel that fusion electric stations can be designed for ever lower levels of emission, to the point at which the additional radioactive load threatened by nuclear power in the fusion era may become vanishingly small. The prospect of central electric generation by plants that will reject no krypton and will release vanishingly small quantities of tritium identifies the low-level radioactive emissions problem as a cost that is for the most part peculiar to fission reactors. Eventually this cost should yield to significant reduction, if not to absolute elimination, with the development of controlled fusion power.

Reactor Safety: "The Tail of a Dead Dragon"

Basically different from those dangers which derive from persistent low-level releases are threats of catastrophic radioactive releases as a consequence of major reactor accidents.[7]

Only some 300 reactor-years of experience with commercial nuclear stations had been acquired by the American nuclear industry to the mid-1970s. Even the 2000 or so reactor-years of experience, if military reactors are included in the statistical base, hardly provide an adequate empirical ground for probability estimates of future failure rates. Because of the poverty of accumulated empirical information, anticipated problems of working reactors must be appraised to a considerable extent on the basis of design data.

The core of a light-water reactor cannot explode. The material in the fuel elements of a BWR or PWR is not the nearly pure U-235 (or plutonium) needed in a bomb. Unfortunately, chances of a nuclear excursion, though still remote, are greater in the case of a high-temperature gas-cooled reactor, a breeder, or a Canadian heavy-water reactor. These devices, which have higher levels of neutron flux than do light-water reactors, present more challenging engineering difficulties. And since the coolant does not also serve as the moderator in these devices, a loss of coolant will not automatically stop the chain reaction. By contrast, in an LWR not only would a loss of coolant interrupt the chain reaction by removing the moderator, but control rods would automatically drop into the core to capture any remaining neutrons whose continuing bombardment of uranium or plutonium could maintain a nuclear reaction.

Evaluating the "Reference Accident"

Yet some dangers *do* remain: circumvention of safeguards by a maniac; a vessel rupture, permitting leakage of toxic wastes or irradiated cooling fluids to the environment; or simultaneous compensating breakages or "common mode" failures.

In early 1975, at the TVA Browns Ferry plant in Alabama, a maintenance worker searching for an air leak with a 25-cent candle set fire to the insulation of the wiring in the unit's control system. Both primary and emergency circuits were incapacitated. Only timely activation of a secondary pumping system not intended for the purpose prevented a full loss of coolant. Pronuclear and antinuclear lobbyists will long debate the lessons of Browns Ferry. What seems *un*debatable, however, is the fact that so complex a facility as a modern atomic plant can be brought to the edge of catastrophe by the most prosaic and uncunning of interruptive events.

The so-called reference accident would consist of a core meltdown in a 1000-megawatt reactor following failure of both primary and emergency cooling systems, accompanied by a rupture of the containment vessel with discharges of fission products directly to the atomosphere. Much of the analytical effort in the field of reactor safety has been directed to estimating the likely consequences of such an accident.

If pipes in both the primary and emergency cooling systems should suddenly rupture, resulting in a loss of core cooling water, the chain reaction would immediately stop for lack of a moderator. But residual radioactivity in the core would continue to generate heat. Actually, only 93 percent of the heat in a typical light-water reactor comes from the chain reaction. The remainder comes from radioactive decay in the fuel elements—decay that continues even after shutdown. Hence two commentators on reactor safety, Joel Primack and Frank von Hippel, have likened the threat of nuclear accident to "the thrashing tail of a dead dragon."[8]

Unless new coolant immediately gets to the core, the buildup of heat will weaken the cladding of the fuel rods. Gaseous fission products from burst fuel elements will collect within the reinforced concrete dome that serves both as radioactive shield and as containment vessel for a light-water reactor. In the event of an explosion sufficient to rupture the containment vessel, fission products would be released to the atmosphere. Carrying downwind, they could represent a hazard for literally hundreds of miles. Meanwhile—and much more likely than an explosive breaching of the containment vessel—the superhot core could burn downward through the floor, sinking into the earth and contaminating ground waters.

No meltdown has ever occurred. Not yet, at least. Indeed, only one accident has lead to fatalities, at the Atomic Energy Commission's reactor testing laboratory in Idaho, where three workers died in 1961. And only once has radiation in large quantities escaped to the surroundings, at Windscale, England in 1957.

Nevertheless, as more reactors have been brought into service, with still more being planned each year, concerns over nuclear plant safety have increased rather than decreased with time. Responding to rising levels of apprehension stimulated by vocal critics from the scientific community, in mid-1972 Atomic Energy Commission officials contracted for a multimillion dollar study of light-water reactor safety. This study, directed by Norman Rasmussen, Professor of Nuclear Engineering at M.I.T., was aimed not at denying the existence of a hazard but at estimating the threat actually posed by commercial fission reactors of current design. The report of the Rasmussen group suggested that nuclear plants represent measurably less of a threat to public safety than do a variety of other hazards to which members of the general population regularly expose themselves, such as death in traffic or by drowning; that in an economy with 100 working nuclear plants, the number of fatalities from a nuclear accident in any year would fall significantly below the number caused by fires, air crashes, or dam failures; that from the most serious catastrophic nuclear accidents, in which a thousand persons might be killed, members of the general population are considerably safer than they are from meteor crashes![9]

Yet, the Rasmussen study showed that reactors are far from proof against failure. Twenty-three hundred immediate deaths could occur in the worst case considered to be feasible, *plus* more than twice that number of induced cancers, many of them eventually fatal, *plus* upwards of $6 billion in property damage. These findings underscore the need both to find safer nuclear plant configurations and to educate downwind communities in techniques of evacuation if an accident were to occur. As will be discussed in Chapter 5, grouping reactors in power parks sited on America's intermetropolitan frontiers can help to meet these needs.

Racing the Probabilities

Drafts of the Rasmussen report began to circulate through the nuclear community in 1974. A subsequent study of reactor safety, conducted under the auspices of the American Physical Society, uncovered significant flaws in the Rasmussen methodology.[10]

Rasmussen's study group had not considered all the long-term effects of residual airborne radioactivity following a nuclear accident. Nor was adequate consideration given to the effects of released radioactivity as a cause of operable tumors, such as thyroid nodules. Taking such factors into account, the American Physical Society team increased the estimate of fatal cancers following a major accident to upwards of 10,000. (This figure may usefully be compared with the more than a quarter million deaths annually from cancers of all kinds in the United States in the mid-1970s.) A core meltdown in a 1000-megawatt plant, it now seemed, would on the *average* result in a higher population exposure to radioactivity by a factor of three than the Rasmussen group estimated for the *worst* case.

Whether the danger to populations is held to exist at the lower (Rasmussen) or higher (American Physical Society) level, most students of the safety issue agree that the likelihood of a major accident, such as a meltdown, is small. But such an accident *will* occur, sooner or later, if enough reactor-years of fission plant operation are accumulated. The appropriate response is, first, to make sure that such an occurrence does not take place sooner; and second, to try forestalling a later occurrence as well by completing the shift away from fission before the odds catch up with the nuclear industry.

Holding light-water reactor commissionings between 1980 and the year 2000 to the lowest possible level offers a reasonable chance to begin replacing fission plants with fusion reactors before the industry reaches 10,000 reactor-years of operation—the point by which, on Rasmussen's showing, a truly catastrophic reference accident, causing 1000 fatalities, will probably occur. It may be possible to prevent that accident from occurring, at least for a while, by drawing on extraordinary reserves of care in the operation of nuclear plants. The "business as usual" attitude has no place in reactor operations. And even in a capital-tight economy, extra investment in safety systems seems fair enough to demand on a temporary basis.

But it hardly seems reasonable, or even responsible, to assume that commitments to extraordinary care and costly safety systems can be honored in perpetuity. Moreover, the cadre of nuclear safety engineers trained in the tradition of Hyman Rickover—the curmudgeon who never flagged in his advocacy of nuclear development, but who also insisted on the strictest of quality control and managerial efficiency—will soon pass from dominance in the industry. As atomic power not only grows but becomes fully commercialized, the quality of program managers may be diluted and the incentives distorted toward profits at the cost of safety.

Nuclear Safeguards: Managing Radioactive Wastes

The draft Rasmussen report appeared in mid-1974, as plans for plutonium recycling began to take shape within the Atomic Energy Commission. The controversy over nuclear power shifted from a focus on reactor *safety,* to the hitherto rather neglected issue of *safeguards.*[11] Safeguards refer to two interdependent problems: the segregation of plutonium and radioactive wastes from populations and from the environment; and special efforts to secure bomb-grade nuclear materials from diversion to unwarranted uses.

High-Level Wastes

Burn-up of uranium in a nuclear reactor produces 300 or more radioactive substances in addition to the already considered krypton and tritium. The most hazardous of these radioactive substances—the so-called high-level wastes—call for permanent disposal and segregation from the environment; for reprocessing to recover plutonium and any unspent uranium; or in some cases, as we shall see, for recycling in order to keep them in the active inventory of the industry.

Actinides: The "Transuranics." The most longlived residues of nuclear burn-up include actinium, neptunium, plutonium, and the other so-called transuranics.

These elements, termed "actinides" after the first element in the series, have atomic weights heavier than that of uranium.

Actinides also have surpassingly long radioactive half-lives. Plutonium has a 24,600-year half-life, and neptunium lasts much longer than that. Moreover, the amount of plutonium in a nuclear waste storage cask will actually increase during the first several thousands of years as a consequence of the radioactive decay of heavier isotopes. Thus a 50-curie deposit of plutonium-239 will more than double in amount before 10,000 years of storage have been completed.[12]

Owing to the extraordinary toxicity of the actinides, these elements remain a threat to organisms long after decay has reduced them to mere trace quantities. As a result, secure management of transuranics presents a million-year problem for mankind—the time span beyond which, it has been computed, the out-load of a typical LWR will have reduced itself through radioactive decay to relative harmlessness.

Fission Products. The other main products of a light-water reactor have much shorter half-lives than do the actinides. But the toxicity of these, the so-called fission products, coupled with their liability to incorporation in human tissue make them frightfully hazardous. This category includes radioactive iodine, strontium, and cesium. Iodine, which concentrates in the thyroid gland, would pose the severest threat in the event of a postmeltdown release of reactor products to the atmosphere. But the brief half-life of this nuclide—about 8 days for I-131 and 20 hours for the more abundant I-133—reduces its persistency in food chains.

It is otherwise with cesium-137 and strontium-90. These fission products have half-lives of near 30 years. But even small quantities of these isotopes pose dangers over long periods of their decay chains—for several hundreds of years.

Ten years after outloading, the wastes of a 1000-megawatt reactor will contain more than 8.5 million curies in cesium and strontium, compared with the 50 curies of plutonium in the same load. After 100 years, the Ce-137 and Sr-90 burden will have decayed to about 100,000 curies; and after 1000 years, to less than a thousandth of a curie.[13] Therefore, these fission productions are commonly thought of as posing a thousand-year disposal problem.

Since both the million-year actinides and the thousand-year fission products are persistent as well as potentially lethal, a strategy of dispersal or dilution is unacceptable. Yet there is no basis in experience for knowing whether any available container will last more than a fraction of the toxic life of the material it stores.

Indeed, the record to date can hardly be thought reassuring. By April 1969, more than 75 million gallons of high-level wastes had been stored in special tanks at various locations in the United States. Fifteen known failures had occurred, eleven of them at Hanford, Washington, where 150,000 gallons of wastes known to have high concentrations of cesium-137 leaked out and began decaying.

Breaching of containers does not necessarily imply widespread contamination. Wastes can remain immobile for literally centuries in a dry, stable underground environment. Therefore, fears center on the possibility of further seepage at Hanford to connected ecosystems, as by an unexpected shift in ground waters.[14] Even nuclear advocates admit the need for better concentrative technology to improve the controllability of radwastes, and for better storage techniques to ensure permanent decoupling of disposal sites from the ecocycles. Since Hanford, major progress has been made on radwaste solidification and concentration.

Yet more than three decades after start-up of the first atomic pile by Fermi, more than 15 years after America's first commercial light-water reactor came on line, leaders of the nuclear community had failed to devise a generally acceptable final plan for long-term management of high-level wastes.

In the late 1960s, Atomic Energy Commission officials staked their hopes on a plan whereby wastes would be trucked to a "remote site" near the town of Lyons, Kansas for permanent storage in geologically stable underground salt beds. Alas, however, after the plan had been announced, studies by the State of Kansas' Chief Geologist suggested previously unappreciated vulnerabilities. The "stable" geology of the state had been altered—not by natural events, but by human intervention—to the point of rendering the salt beds unreliable repositories. Generations of oil drilling and solution salt mining had weakened the strata in which the waste caskets were to be entombed.

Moreover, the citizens of Lyons drew the logical conclusions from Washington bureaucrats' repeated emphasis that their town was next to nowhere. Rhetoric that had reassured Americans in Maine and California only frightened the Kansans. Political opposition in Kansas, where the site was not considered "remote," contributed to the reversal of the Commission plan.

The collapse of the Lyons salt-mine storage scheme left the AEC's successor agency, the U.S. Energy Research and Development Agency—and indeed left American society generally—with a serious and still unresolved problem.

Figure 3-1 suggests the approach espoused by M.I.T.'s David Rose, an internationally regarded nuclear engineer. Rose's plan, devised in collaboration with A.J. Kubo, calls for the special separation of actinides from fission products.[15] The actinides would be clad in fuel elements for recycling back into reactors. Ideally, the cores of the reactors into which they would be inserted could render the radwastes less hazardous through nuclear transmutation. Neutron bombardment in working cores should eventually transmute most of these actinides into less dangerous isotopes, just as transmutation breeds plutonium from uranium in a reactor to begin with. On the average, Rose has computed, the products of such transmutational processes will need secure storage for 700 years. In this way, Rose and Kubo propose to eliminate the million-year problem—or at least, to change it into a 1000-year problem.

In yet another special filtration process, radioactive cesium and strontium would be separated from the remaining fission products. In the 1950s and

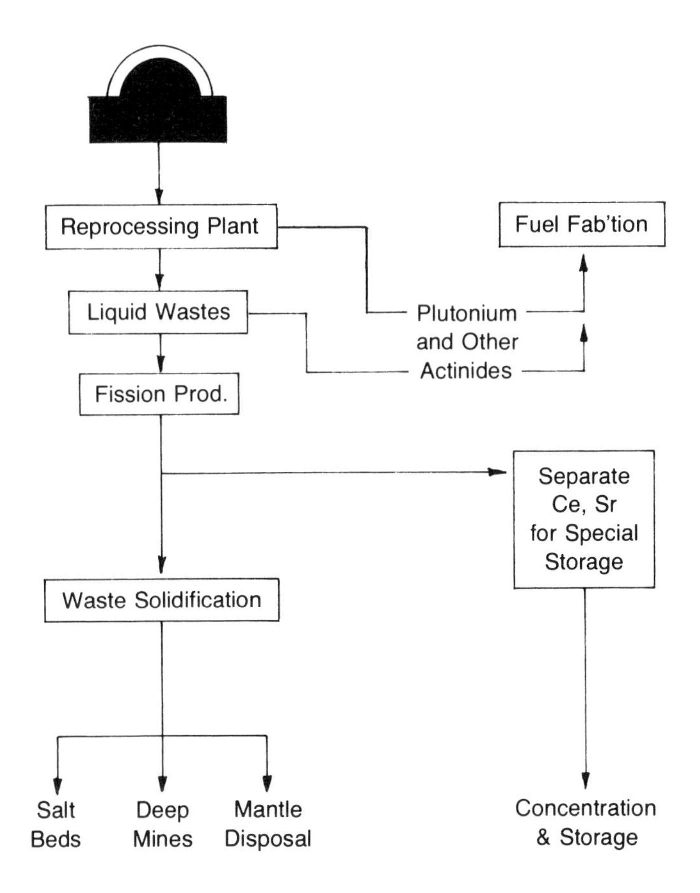

Figure 3-1. Separation and Disposal of Radioactive Wastes.

Notes to Figure 3-1. Safe disposal of highly irradiated reactor wastes may hinge on three processes: (1) a capacity for the discriminating separation and "sorting" of different fission products; (2) recycling of unstorable wastes back into reactors for burn-up; and (3) careful long-term storage of remaining radwastes.

In a series of separative processes, materials that may be used as fissionable fuels (such as plutonium) or that pose exceedingly forbidding long-term disposal difficulties (other highly toxic, long-lived actinides) are first removed from reactor wastes. These are packed into new fuel assemblies and cycled back into reactors. In this way, such materials will never invite the illusions of "safe storage." Instead, they will be carried in the active fuel inventory of the nuclear industry until further neutron bombardment within reactor cores destroys them. Over time, recycled actinides will themselves be burned up. That is, they may be converted into fission products of shorter half-life and reduced toxicity.

Following the removal of plutonium and other actinides from spent fuel elements, the Rose-Kubo scheme calls for a further separative process to remove such hazardous elements as radioactive strontium and cesium for special concentration.

In the Rose-Kubo plan, any remaining fission products are then solidified, preparatory to disposal in long-term storage facilities. Attention to the disposal problem in the 1970s focused ever more intensively on the identification of acceptable long-term disposal methods. Salt bed, deep mine storage, and mantle disposal represent only three of the possibilities under study. Once sound methods for long-term disposal are found, areas suitable to this function could be given special attention as potential sites for power parks.

1960s, as components of bomb-fallout in the atmosphere, these nuclides entered the muscles and bones of children by way of the grass-cow-milk food chain. The human body, mistaking cesium for sodium and strontium for calcium, incorporates them into tissue. There they undergo radioactive decay, acting as known carcinogens. But cesium and strontium are also readily concentrated. Were it otherwise, these isotopes would be less inclined to collect in the bodies of animals and children. Their liability to concentration means that following separation, they should be segregated with relative ease from the environment in special storage facilities.

Any remaining fission products could be solidified and compacted, preparatory to deep burial in salt beds, in special mines, or even at the interfaces of the tectonic plates on which the continents drift, where the wastes would slowly be drawn by geological movements beneath the mantle of the earth.

The need to test and validate schemes such as Rose's must take first priority in the technical research program of the nuclear industry. Rose has stressed the urgency of research on radwaste disposal. Behind the exhortation is recognition that Atomic Energy Commission officials throughout the 1960s concentrated on glamorous reactor development programs (e.g., the breeder), while financially starving other essential efforts.

Low-Level Wastes

"Low-level wastes" refer mainly to contaminated materials whose radioactive contents are distributed in dilute form. Such wastes contain the prosaic residues of the nuclear industry: cleaning cloths, discarded worker garments, and the like.

As with mine wastes and mill tailings, storage of low-level residues from reactors can present a serious volumetric problem. The main challenges to effective disposal have in the past centered on the acquisition of commercial space for disposal and economical transportation.

Low-level wastes from fuel fabrication plants and reactor and reprocessing facilities contain not only relatively weak and short-lived contaminants, analogous to radon, but also long-lived transuranics. Toxicity per unit of bulk is low. But the bulk itself is formidably large—which means that the *absolute* amount of potentially dangerous radiation may also be considerable. Reliance on Rose-Kubo procedures for recycling and special deep disposal of *high-level* wastes would not, therefore, fully solve the million-year problem—unless provision also were made to purge the environment of the actinides that exist in dilute form throughout *low-level* wastes. Is such a plan for special separative and disposal processes, directed at low-level radwastes, practicable?

At this point, economic considerations emerge as serious constraints on the nuclear industry's ability to handle low-level wastes in this way. The transuranics exist in such dilute form that prohibitive volumes would have to be processed in order to concentrate all actinides for appropriate treatment. The nuclear cost picture could be skewed beyond the margin of profitability if the fuel cycle had to bear the cost of such an elaborate waste treatment system.

The alternative to costly purging of dilute actinides from low-level wastes is a continuation of current storage practices: simple underground burial or long-term deposition in commercial loft space. The prime requisite remains that of immobilizing the dangerous components where they are sure to be undisturbed. John Price, in his study of nuclear energetic balances, has suggested that the energy required to maintain safe storage facilities for the transuranic wastes in these low-level residues could equal or even exceed the total useful power output of the fission plants that produce the long-lived wastes in question.[16] Price's estimates assume that million-year storage will be required. His conclusion, then, might be offset in either direction, depending on the assumptions made regarding the needed duration of the waste management facilities.

If cheaper waste disposal techniques than Price considered were devised, or if the storage time requirement were reduced, the energy balance could tip to show a more favorable verdict for fission power.

The crucial point seems to be that the million-year problem can perhaps be brought to manageable proportions. But it probably will never be eliminated. Indeed, paradoxically, the problem occurs in its least tractable form in connection with the so-called low-level wastes rather than in connection with the high levels, where a plan such as Rose and Kubo's may work.

Any solution to the low-level waste problem—and such a solution can at best be imperfect or partial—will probably require that fuel fabrication plants, reactors, and disposal areas be configured in ways likely to facilitate control and segregation. Such requirements point toward the desirability of minimizing the number of points at which low-level wastes are produced, and of reducing the distances between these points and bulk waste depositories. Planning to cluster reactors and support facilities in power parks and then siting nuclear parks in geologically stable, underpopulated areas of the American midcontinent could contribute to a responsible approach.

Summary: The Radiological Case against the Atom

Nuclear power is not—and it probably never can be—perfectly safe. *How unsafe is it?*

Limitations of experience with reactor operations as well as of radiological knowledge prevent precise estimates of the nuclear risk. However, it seems possible to assert in general terms that atomic power can be brought to a level of safety that compares favorably with fossil-fired electricity. The estimates presented in Table 2-1, for example, suggest that the costs of air pollution and occupational hazards associated with fossil plants exceed the hazards (including the hazards of nuclear waste management) of atomic electric generation by almost a factor of four. Even substantial allowance for error in these estimates leaves room for claims of the superiority of the nuclear form in terms of health effects.

The assignment of the safety edge to atomic power does not, however, imply that a wholesale switch to the nuclear form would leave no remaining risks. Politically and socially, the implication is that specific regions of the country must be found whose residents will knowingly absorb some of those remaining risks, in consideration of certain economic advantages that would accrue as a consequence of their acceptance of nuclear development in the region.

While the more worrisome implications of nuclear power reach beyond the margin of the present generation as well as beyond the boundary of any particular region in which reactors may be sited, even the staunchest nuclear advocates recognize that lapses in safety or safeguards procedures could have catastrophic effects, at least on local populations and on the regional ecosystem.

The resulting challenges are twofold. First, through better engineering and improved reactor siting concepts, it is necessary to bring the risks of fission plants down to the lowest level consistent with reasonable economic constraints. A strategy that combines a transition from fission to fusion with the adoption of clustered siting of reactors in nuclear centers may promise a satisfactory response to this first challenge.

Second, it is necessary to identify regions whose citizens will accept the remaining risks of fission power. As will be discussed in Chapter 7, in the course of satisfying this requisite, populations in specific regions may conclude a series of "interregional nuclear bargains." By these bargains, host regions would gain the benefits (and accept a disproportionate share of the risks) of long-term nuclear development. Other regions would forego the economic stimulation of nuclear center construction within their boundaries. But their citizens would gain the security of knowing that sites are available elsewhere (i.e., in host regions) for the reactors that are needed to supply the nations' electricity in the twenty-first century.

A singular incentive for regional leaders to accept large-scale fission development derives from the right thereby gained for the same regions to profit from the eventual switchover to controlled thermonuclear fusion. Plainly, such an approach rests on the assumption that fusion power will indeed materialize toward the turn of the century. To a consideration of the prospects for such a switchover we must now turn.

Notes

1. Lester Lave and James Freeburg, 14 *Nuclear Safety* (September-October 1973), 423.

2. D.D. Comey, "The Legacy of Uranium Tailings," 31 *Bulletin of the Atomic Scientists* (September 1975), 43, citing unpublished computations by R.O. Pohl.

3. See Hans Bethe's summary article, "The Necessity of Fission Power," *Scientific American* (January 1976), 21, pp. 23-25.

4. The basic figures on krypton and tritium effects come from my *Energy, Economy, Ecology* (New York: Norton, 1972), pp. 138-144. I have taken additional data on tritium from unpublished reports by analysts at Battelle Pacific Northwest Laboratory who are developing the generic impact environmental statement on fusion electric systems.

5. See the discussion referenced by note 23, Chapter 5.

6. See, however, the views of a prominent antinuclear scientist, Ernest J. Sternglass, in *Low Level Radiation* (New York: Ballantine, 1972), especially Chapter 14, on the alleged radiological threat of civilian reactors.

7. Physicist Ralph Lapp has published several popularized analyses of the nuclear disaster potential. See especially "Safety," *New Republic* (January 23, 1971), 18, and "The Four Big Fears About Nuclear Power," *New York Times Magazine* (February 7, 1971), 16.

8. Joel Primack and Frank von Hippel, "Nuclear Reactor Safety," 30 *Bulletin of the Atomic Scientist* (October 1974), 5, p. 7.

9. *An Assessment of Accident Risks in U.S. Commercial Nuclear Power Plants* (U.S. AEC, WASH-1400, August 1974). See for an excellent short summary and commentary, Robert Gillette, "Nuclear Safety: Calculating the Odds of Disaster," 185 *Science* (September 6, 1974), 838.

10. The American Physical Society review appeared in Supplement I, 47 *Reviews of Modern Physics* (July-August 1975), pp. 51ff.

11. See The Union of Concerned Scientists', *The Nuclear Fuel Cycle: A Survey of the Public Health, Environmental and National Security Effects of Nuclear Power* (Cambridge, Mass.: Colonial Press of the MIT Press, Revised edition, 1975), especially Chaps. 4, 6, 8.

12. Tables 3 and 4 of John Holdren's "Hazards of the Nuclear Fuel Cycle," 30 *Bulletin of the Atomic Scientists* (October 1974), 14.

13. Ibid.

14. On the Hanford seepage, see Robert Gillette, "The Anatomy of an Accident," 181 *Science* (August 24, 1973), 728.

15. A.S. Kubo and D.J. Rose, "Disposal of Nuclear Wastes," 182 *Science* (December 21, 1973), 12-5-1211. See also in general on the radwaste disposal problem, *Siting of Fuel Reprocessing Plants and Waste Management Facilities* (Oak Ridge National Laboratory, ORNL-4451, July 1970). Also of value are the comments by James Liverman and David Rose before the Subcommittee on Energy and the Environment of the House of Representatives, Part I, Serial no. 94-16, 94th Cong., 1st, April 29, 1975, pp. 541ff., and especially pp. 572-577;

and "A Real Lag in Developing Atomic Power? . . . Interview with William A. Anders," *U.S. News & World Report* (September 8, 1975), p. 31.

16. "Dynamic Energy Analysis and Nuclear Power" (London: Press-on-Printers, a Friends of the Earth report, December 18, 1974), p. 23.

4 Fission to Fusion: The Promethean Progression

An H-bomb explosion suggests the magnitude of the destructive power of an uncontrolled thermonuclear reaction. Within the energy community, CTR means "controlled thermonuclear reactor." *Control* refers to a capability—yet to be achieved—for regulated releases of energy that may be channeled into some useful form, such as the movement of turbines to generate electricity. By the further requirement of *economy,* a fusion device must be made to release more controllable energy than is required to produce the reaction in the first place. The thermonuclear devices that have succeeded to date by the test of economy—that is, H-bombs—fail by the test of controllability.

But major gains registered in thermonuclear research between 1965 and 1975 suggest that the fusion scientists will indeed soon enough have truly harnessed the fires of the second sun. So promising was progress during this decade that in the early 1970s fusion power began to appear as a serious competitor of the breeder reactor. Even as the LMFBR program began to experience its series of compounded technical and economic difficulties, scientists at national laboratories of the Atomic Energy Commission (especially at Livermore, California; Los Alamos, New Mexico; Oak Ridge, Tennessee; and at the Princeton University Plasma Physics Laboratory) began to achieve fundamental breakthroughs. In 1971, no less a figure than Hannes Alfven, the Nobel laureate physicist leader of antifission lobbyists, wrote:

> In my opinion, a solution of the fusion problem is less distant today than the Moon was when the Apollo project got started. This means that if a national effort of the same kind as the Apollo program were made, fusion energy would be available in a comparable time.[1]

The U.S. Energy Research and Development Administration's "National Plan," released on April 15, 1976, calls for an operational test fusion electric system in the late 1980s. In the same document, it is admitted that a final decision on the commercial acceptability of the LMFBR cannot be expected before 1986. The ERDA timetable, which makes a breeder bypass seem feasible despite strong support for the LMFBR within the Ford administration, suggests construction of a demonstration commercial fusion plant around 1995.[2] Table 4-1 reflects a scenario in which (1) reliance on the light-water reactor is stretched out through plutonium recycling and use of lower-grade uranium ores; (2) the LMFBR is bypassed; and (3) a major effort is launched to meet the LMFBR's

Table 4-1
Increasing Share of Electric Generation by Fusion Power, 2000-2020

	Assumed Electric Output, Thousands of Mwe from Indicated Source in Year		
	1980	*2000*	*2020*
Fission-Based			
Light Water Reactors			
U-235 Burning	80	650	400
Pu Burning from Fusion Hybrid		160	170
(Hybrid Deduction)		(−10)	(−10)
LMFBR			
Fusion-Based			
Fusion Electric Plants		1	1090
(Fusion Torch Deduct'n)			(−50)
All Other	420	800	1200
	500	1601	2800

share of the nation's electric demand in the twenty-first century with fusion plants. The capacity buildup rates implicit in this scenario are shown in Figure 4-1.

(As a stopgap, some plutonium might be bred using special fusion reactors "hybridized" with fission devices. The hybrid concept as well as that of a modified version of the so-called fusion torch—by which thermonuclear reactions might be used to scavenge the wastes of light-water reactors—will be dealt with later.)

The Physics of Plasmas

Thermonuclear reactions occur in superheated gases called plasmas. When raised to several hundreds of thousands of degrees, a gas undergoes basic physical changes, leading to the characterization of plasmas as representing a fourth state of matter. At such temperatures, electrons dissociate from their atomic nuclei. The now-naked nuclei thus become free to interact directly with one another. Plasmas have two further properties: they can carry electric currents, and they can themselves be shaped by magnetic fields.

We shall see in what ways physicists use these unusual properties in their effort to produce controlled thermonuclear fusion.[3]

The first thermonuclear reactors will contain plasmas of deuterium and tritium. In such CTRs, the nuclei of these isotopes of hydrogen would fuse to create helium, plus fast-moving neutrons, plus heat. This, the so-called D-T reaction, will occur at a lower ignition temperature than will any of the other reaction possibilities. In addition, D-T fusion releases a relatively large total quantity of energy—from 17 million to as many as 22 million electron volts per reaction, compared with a release of only 3 to 4 million electron volts for deuterium-deuterium (D-D) fusion.

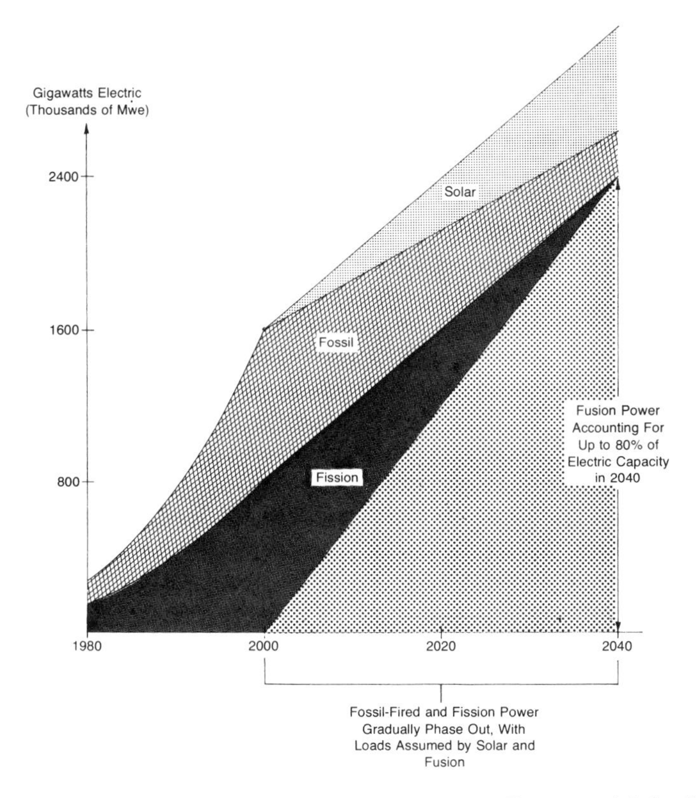

Figure 4-1. Preemption of Electric Production by Fusion and Solar Power.

Notes to Figure 4-1. The curve showing a tenfold increase over 20 years from 80 gigawatts of electric capacity in 1980 roughly corresponds to the likely buildup of installed fission-based nuclear power in the United States. From the year 2000 onward, it is assumed that the buildup of nuclear electric capacity—including fusion power and any breeders, as well as light-water reactors—will proceed linearly at an average rate of 40,000 megawatts per year. Total nuclear capacity in 2020 should then be twice the level in the year 2000.

Fossil-fired electrical generation should decline both relatively and absolutely after the year 2000. And solar power should be brought on line to supplement fusion as a source of clean electric power.

Throughout the entire discussion, it should be remembered that solar power cannot contribute significantly to *central* electric generation of base-load electric power unless breakthroughs are achieved on the cost front. If solar power can economically be used only in limited, decentralized market sectors—such as house heating—then the wedge at the top of the figure will be much thinner than is shown. In this case, the buildup of fusion power after the year 2000 may have to proceed even more swiftly.

By 2020, fusion should have begun to assume a major and increasing share of the nation's generating load, with light-water reactors and any breeders accounting for a correspondingly smaller fraction. For purposes of preliminary estimation, assume that fusion installation can occur at a linear rate significantly higher than today may be achieved with fission reactors. Assume too that fusion is desired to overtake, and thereafter preempt, the entire American nuclear program in the year 2040.

In this preemptive scenario, fusion would grow from 0 to 1200 gigawatts in the 20 years after its assumed commercial introduction in the year 2000. The equation of fusion buildup would intersect the total nuclear buildup line in 2040 at the 2400 gigawatt level of capacity.

A disadvantage of D-T as compared with D-D fusion lies in the fact that the needed tritium is virtually unknown in nature. Because it is radioactive, tritium also presents certain handling problems, both in its creation as a fuel (it is bred from the relatively abundant element lithium), and in the disposal of tritium wastes during the operation of a fusion reactor. Fusion scientists seem confident that the problems of tritium management are solvable—a familiar mood of self-confidence among nuclear physicists, and one that should perhaps put less involved observers of the energy scene on guard.

Both deuterium and tritium carry positive electrical charges. Hence they repel each other, and with a force that increases in proportion to the proximity of the nuclei. In order to get a nucleus of deuterium to collide with a tritium nucleus so they may fuse, the two must be hurled at each other at enormous speeds. Speeds high enough to overcome the repellent forces of the positive electrical charges correspond, in thermodynamic terms, to the imparting of high temperatures to the D-T plasma. These temperatures must exceed tens of millions of degrees.

No solid material—no alloy of steel, no exotic ceramic—can reliably confine reactions occurring at the temperatures of thermonuclear reactions. As soon as a hot deuterium or tritium nucleus should hit the side of a solid containment vessel, it would immediately lose most of its energy—that is, it would cool off and hence slow down. A "fusionable" plasma cannot survive in such circumstances. Hence a solid confinement chamber might defeat the very purpose of the reactor of which it would be a part, for its walls would extinguish the nuclear fire.

The most promising response to the problem of plasma confinement turns on the susceptibility of ionized gases to control by electromagnetic fields. A D-T plasma may be shaped, and hence confined, within a properly designed "bottle" consisting of magnetic lines of force. As shown in Figure 4-2, a donut-shaped device called a *tokamak,* from the Russian abbreviation for toroidal magnetic chamber, is wound with wires. When carrying an electric current, these wires generate an electromagnetic field. The field builds an invisible torus within the machine—the electromagnetic bottle—which then actually holds the superheated D-T plasma.

The need to ignite the D-T plasma and to create the magnetic bottle for its confinement accounts for the huge inputs of energy that are needed to get a fusion reaction. To satisfy the condition of economy in the energy balance of a CTR, the task is simultaneously to attain conditions of (1) plasma temperature, (2) plasma density, and (3) plasma confinement that ensure that more useful energy will result from the fusion reactions than must be put into the reactor to begin with.

Plasma physicists believe that in tokamak-type systems, the fantastic temperatures needed for ignition may be attained by running a current through the plasma. Such a current will generate resistance heat on much the same principle that is used in an electric toaster. The plasma may be further heated by rapidly compressing it within the electromagnetic force lines, thus "pumping" it as if by closing the piston in an invisible cylinder.

Source: R.G. Mills, et al., *A Fusion Power Plant* (Princeton University: Plasma Physics Laboratory, August 1974), cover illustration and pp. 10, 16, 148, 513, 520. Reprinted with permission.

Figure 4-2. Artist's Conception of a Tokamak Electric Plant.

Notes to Figure 4-2. Fusion research has progressed to a point at which it becomes necessary to begin considering the practical problems of engineering a controlled thermonuclear reactor (CTR) for electric generation.

A tokamak-type plant, of the sort under conceptual study at the Princeton Plasma Physics Laboratory, would represent a huge and enormously complex project. Within the donut-shaped torus shown in this artist's conception, neutrons of high kinetic energy would be produced by deuterium-tritium reactions. The energy of these fast neutrons, along with any heat directly released by plasma reactions, would then be used to raise a head of steam for electric generation.

The deuterium could be taken cheaply from water. But because tritium (T) is virtually nonexistent in nature, lithium (Li) within a blanket of molten salt inside the toroidal vacuum chamber would be used to breed tritium, following the reaction given by:

$$\mathrm{Li} + 1\,\textit{neutron} = \mathrm{He} + \mathrm{T}$$

In some cases, this reaction–of which helium (He) is the main chemical product–also yields another neutron. This "extra" neutron, by colliding with still another lithium nucleus, then breeds even more tritium.

The illustrated plant (rated at 5305 megawatts thermal) would cover up to 75 acres of land. Fewer than 400 of the plant's 2030 megawatt electrical output would be needed for internal functions–mostly to drive helium circulators in the cooling system. Mid-1974 estimates suggested that with support facilities, the plant would cost an estimated $1.2 billion to bring on line and could then deliver electricity for 15.17 mills per kilowatt-hour.

(Work also goes forward on a radically different technology for confining and heating plasmas. In the inertial confinement system being developed at the Los Alamos and Lawrence-Livermore Laboratories, a tiny pellet of deuterium mixed with tritium would be compressed by a collossal implosion triggered by laser beams, as in Figure 4-3. Experimentation is also proceeding on the use of concentrated ion–or electron–beams to drive the implosion. With laser fusion, the forces of inertia rather than an electromagnetic field would confine the plasma for the billionth of a second during which the laser pulses slam into the D-T pellet, heating it and thereby causing fusion.)

The plasma must be sufficiently dense to ensure that enough deuterium and tritium nuclei are in proximity to one another so reactions will occur in adequate numbers. The denser the plasma, the greater the number of nuclear collisions. On the graph in Figure 4-4 the various combinations of plasma temperature, density, and confinement time that satisfy the conditions of an economical energy input-output ratio appear as curves or surfaces, called "break-even" frontiers.

In 1948, a British physicist, J.D. Lawson, defined the critical break-even frontier–that range of temperature, density, and confinement time needed to produce more energetic output from a CTR than must go into the reaction in order to achieve ignition and confine the plasma. The oval-shaped Lawson break-even frontier in the upper left corner of Figure 4-4 applies to the D-T reaction. Subsequent calculations at the Princeton Plasma Physics Laboratory, associated with the names of John Dawson, Fred Tenney, and Harold Furth, suggested that D-T fusion in a properly configured reactor can effectively occur at the somewhat less restrictive conditions of temperature, density, and confinement time indicated in the break-even frontier shown to the left of the Lawson boundary. Hopes for early success using the Dawson-Tenney-Furth technique hinge on efforts to boost the average velocity of deuterium and tritium nuclei within a tokamak by injecting a special beam of fast-moving neutral particles into the already superheated plasmas.[4]

The leftward dots in Figure 4-4 indicate conditions already achieved by various experimental fusion machines.[5] Constant progress within the fusion effort as a whole toward break-even conditions suggests a successful laboratory experiment, one that demonstrates the feasibility of controlled thermonuclear power, before 1985.

The Vaccine of Confidence

The characteristic outlook of fusion researchers bears brief consideration. Most fusion researchers seem to have been successfully innoculated with the vaccine of confidence. As knowledge of plasma properties has grown–and it *has* grown, exponentially, since the mid-1960s–so has the scientists' appreciation of the

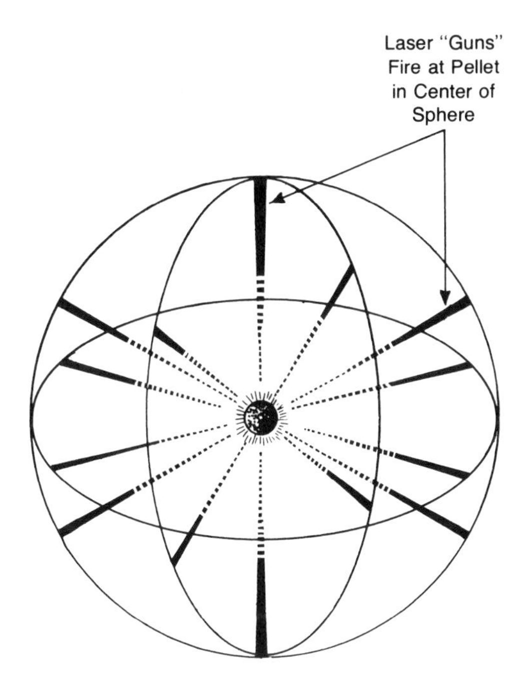

Source: Based on the L.A.S.L. Laser-Fusion Reactor Design, in Division of Controlled Thermonuclear Research, U.S. AEC, "Fusion Power: An Assessment of Ultimate Potential" (February 1973: U.S. G.P.O.), pp. A-14 through A-16.

Figure 4-3. Conceptualization of the Laser-Fusion Reactor.

Notes to Figure 4-3. In the laser fusion systems under development at the Los Alamos Scientific Laboratory, New Mexico, at Rochester University, New York, and at the Lawrence Livermore Laboratory, California, pellets containing deuterium and tritium will be subjected to powerful pulses from a series of carefully focused lasers. The D-T pellets are themselves *one-fiftieth* of an inch or so in radius. The "firing" of these lasers occurs simultaneously and so quickly–in less than a billionth of a second–that inertial forces keep the pellet stably positioned as the impulses hit it with a perfectly balanced effect. Hence such reactors are term ICTRs–inertial confinement thermonuclear reactors.

The implosion created by the laser pulses compresses the D-T nuclei within the pellet. If compression occurs with sufficient speed and impact, the ignition conditions for thermonuclear fusion will be satisfied. In the ICTR as with tokamak-type CTRs of the sort shown in Figure 4-2, ignition occurs only when the plasma reaches a temperature in the range of 100 million degrees centigrade.

The thermal yield from such a reaction, however, will be much less than that of a tokamak-type CTR. With a nominal output of about 200 megawatts thermal for an ICTR of the type illustrated, a series of machines–10 or more–would be "ganged together" to form a single power unit.

The heat from D-T fusion within an ICTR's central cavity would then be borne off by flowing lithium raised to about 750 degrees. This lithium would transfer about half of its energy to water in a heat exchanger and then recirculate. The water, thus raised to steam, would drive turbines as in any other kind of power plant.

Meanwhile, a new D-T pellet would be inserted into the cavity at the focal point of the laser beams.

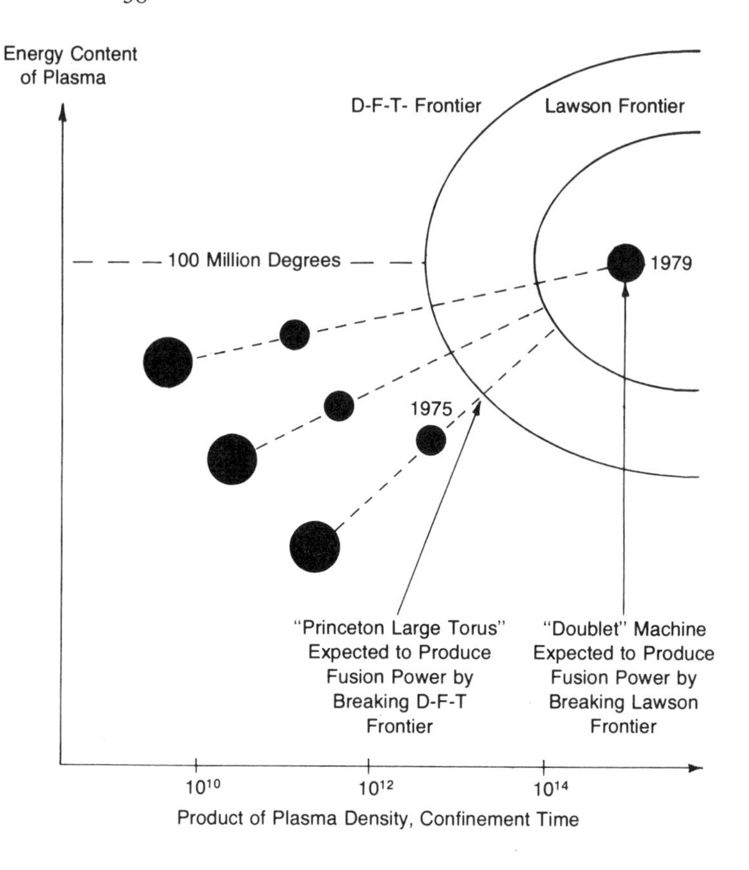

Source: Adapted from Robert Hirsch and William Rice, "Fusion Power and the Environment," U.S. Energy Research and Development Agency, Figure 10, "Break-Even Plasma Conditions for Fusion Power," p. 20.

Figure 4-4. Progress Toward a Controlled Fusion Electric Plant.

Notes to Figure 4-4. Research toward commercially usable fusion electric power proceeds through a series of parallel development efforts. The solid dots indicate approximate conditions of temperature, plasma density, and confinement time which have been attained at different fusion experimentation centers. As each research team improves its devices, the controlled thermonuclear effort advances on several fronts.

The dots, from upper left to lower right, signify the efforts at the Lawrence Livermore Laboratory in California (so-called 2-X mirror machines), at the Los Alamos Scientific Laboratory in New Mexico (the Scylla and Scyllac machines), and at the Princeton Plasma Physics Laboratory in New Jersey. Any one of these projects could produce satisfactory results by the late 1980s.

Tokamak research based at the Princeton Plasma Physics Laboratory should show fusion feasibility with a new machine, the "Princeton Large Torus," in the late 1970s or so. As indicated in the bottom arrow, such an achievement would satisfy the Dawson-Furth-Tenney requirements. Beyond this break-even level, more useful energy will be produced (both in the form of heat and in the form of the kinetic energy of fast-moving neutrons) than must be provided in order to start the reaction. And the so-called Doublet reactor under development by the General Atomic Corporation is expected to produce controlled thermonuclear reactions satisfying the original Lawson ignition condition.

fantastically complex engineering problems posed by the idea of commercial fusion. In post-Great Society, post-Viet Nam America, increasingly afflicted with anxiety and doubt, fusion researchers retain confidence in the face of the most formidable challenges to man's practical ingenuity. The optimism within the fusion research community may represent a resource just as necessary for a response to America's energy problems as are the nation's raw fuel reserves.

Plasmas reveal phenomena technically known as "higher-order effects." For example, the passage of an electric current through a plasma to heat it (the desired first-order effect) itself tends to generate a weak electromagnetic field. This field can compete with the lines of electromagnetic force generated by the toroidal windings. This competitive field interferes with the flow of the plasmas (an undesirable second-order effect). To correct for this minor distortion, the shape of the main electromagnetic containment field must be altered, perhaps by changing the shape of the donut-shaped vessel itself. Such corrective tinkerings in turn may alter the efficiency with which tritium can be bred from the lithium in the vacuum chamber (a third-order effect). And so on.

Each correction induces unforeseen higher-order effects, and each higher-order effect calls for still further correction. From the very first, fusion research has opened a series of linked complexities. The ascending order of subtlety and difficulty in these complexities has thus far been matched by the confidence as well as the ingenuity of the investigators. Perhaps we can never gain the benefits of controlled fusion power without the experimental stamina of the plasma scientists. Perhaps the vaccine of confidence suggests a requirement of more general application in a society whose problems may be as much psychological as they are technological.

Special Uses of Controlled Fusion Power

Some thermonuclear proponents argue that we need not actually cross the break-even frontiers in order to gain benefits from the CTR concept. Fusion reactors, these advocates contend, should be considered under two special configurations that fall short of their use as prime movers of electric generators: as hybridized components of fusion-fission power systems; and as "fusion torches" used in national programs of environmental cleanup.

The Fusion-Fission Hybrid

As rising real costs for uranium in the early twentieth century threaten to inhibit the growth of light-water reactor power, anticipated shortfalls will intensify pressures to develop and commercialize breeder reactors. The implications for energy policy are formidable: Prospects for a rapid buildup of CTRs after the year 2000 will depend on the intelligence and vigor of fusion development between now and the turn of the century; and on the ability to rely on a rapid rate of fusion buildup may depend on America's chance to bypass the breeder.

Can America's bet on fusion power somehow be hedged? The answer is–Perhaps, through an intermediate strategy that falls between a complete breeder bypass and full reliance on LMFBRs. Such an intermediate strategy would support minimum levels of plutonium breeding, but would do so with the use of special fusion devices rather than liquid-metal breeders on the Clinch River Demonstration Plant model.[6] This strategy has two attractive features: it would furnish a limited plutonium base to help bridge the transition from LWRs to CTRs; and it would not inhibit–it would, indeed, encourage–progress in the general field of fusion technology.

Fusion machines produce neutrons in large numbers as the nuclei of deuterium and tritium join to form helium. Indeed, the fusion process inherently produces the neutron-rich environment whose achievement has long been regarded as the objective of a successful breeder. Short of the achievement of break-even conditions, properly designed fusion machines may be used to produce neutrons for the commercial breeding of fission fuel–Pu-239 from uranium, or U-233 from thorium.

In this envisioned use of fusion technology, a thermonuclear device might complement or, preferably, substitute for the increasingly problematic LMFBR. Ideally, the LMFBR program would be scrapped and fission-based breeders bypassed altogether in favor of an immediate commitment to the CTR option.

Since neutrons are needed to sustain a chain reaction–achieved in a conventional LWR by maintaining a critical mass–the excess neutrons from a fusion reaction can also be reflected into a subcritical mass of fissile elements in a modified LWR in order to feed an otherwise unsustainable fission chain reaction, thereby producing fission-based nuclear power *without* a critical mass in the reactor.

The fusion reaction in Figure 4-5 would be engineered to go forward within a blanket of fertile uranium or thorium. Plutonium bred from U-238, or U-233 from thorium, would be loaded into a nearby fission reactor to produce heat for electricity. More heat would be needed to drive the breeding reaction than the fusing of deuterium-tritium nuclei would release. So some of the electricity would be recycled within this symbiotic system to drive the fusion, producing still more neutrons to maintain the breeding cycle.

Relatively little attention has been given by U.S. policymakers to support for such programs. The commitment to the LMFBR, backed by a heavily funded research establishment (not to mention the prestige and the "face" of prominent figures in the breeder lobby), has undercut any basis of political support for potentially competitive efforts.

Of course, fusion-fission hybrids would themselves show some of the disadvantages of liquid-metal breeders, including the production of bomb-grade plutonium or U-233. There exists, too, the danger that successful hybridization in the early 1990s would reduce the incentive to develop fusion machines able to attain break-even conditions for commercial electric generation. Yet on balance,

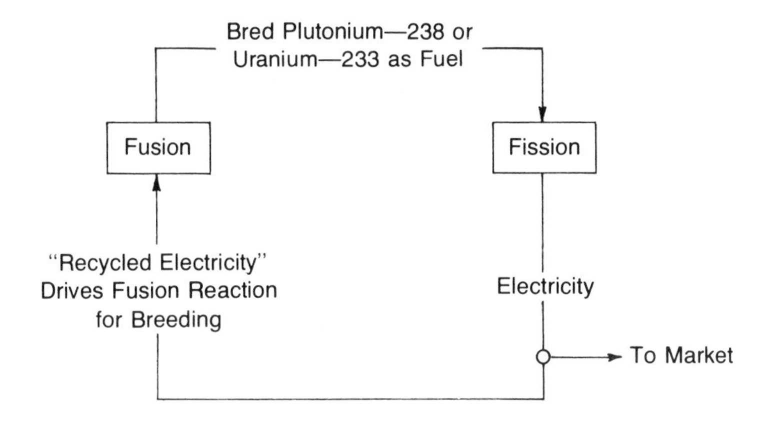

Figure 4-5. Lidsky's Fusion-Fission Hybrid Electric System.

Notes to Figure 4-5. The development of plutonium-burning light-water reactors would invite the introduction of experimental plutonium-breeding fusion-fission systems. In such hybrids, fusion reactions would breed fissionable fuel to be used in LWRs to produce electricity. Some of the produced electricity would then be recycled within the hybrid system to continue driving the breeding reaction.

Lawrence Lidsky of the Massachusetts Institute of Technology is the best-known proponent of hybrid systems. The CTR on the lefthand side of the schematic would breed U-233 from thorium. B.R. Leonard of Battelle Institute has shown that the Lidsky computations imply a doubling time of 10 years.

The fusion component of Lidsky's symbiotic system would require an input of some 565 megawatts of *thermal* energy, and would "pay-back" only approximately 300 megawatts to the system, for a net consumption of about 90 megawatts of *electric* power.

Bred U-233 from the fusion component would be transferred to a co-located fission reactor and there burned to produce electricity. Rated at 4450 megawatts thermal and assuming 40 percent thermodynamic efficiency, Lidsky's hybrid system would produce some 1,690 megawatts of electricity for sale—a figure which allows for the 90 megawatts that must be recycled to the CTR.

In other variants of hybrid systems, a fusion reactor would be physically coupled to the fission device so that neutrons from plasma reactions could directly participate in the chain reaction. Thus the chain reaction might be sustained but in the absence of a critical mass of fissile material within the system as a whole.

if Americans are to develop a plutonium economy of minimal dimensions as insurance against delays in fusion electric progress, it would seem desirable to develop plutonium-breeding devices that logically point to an end of the fission era (as with hybrids) rather than to its perpetuation (as with the LMFBR).

The Fusion Torch

Since hybrid research would proceed as an integral component of the nation's overall fusion effort, the development of symbiotic systems such as Lidsky's could contribute to the perfection of spin-off technologies. One spin-off of fusion research—a more exotic application of thermonuclear power—has been proposed by two plasma physicists, William Gough and Bernard Eastlund.[7]

Pointing to the waste scavenging possibiliites suggested by the fantastically high temperatures at which a fusion reactor can exhaust its plasmas, Gough and Eastlund have argued that CTRs might be used to break down any known physical substance on earth. Fusion research has already produced plasmas at temperatures in excess of 500,000 degrees. Such plasmas might be used to reduce wastes—old cars, appliances, even garbage or sewage—to their elemental atoms. The same principle could also be applied to mining. Of course, the energy needed to drive such a fusion torch may prove far more costly than the scavenging services it could perform would be worth. However, in an era likely to witness a growing emphasis on both the processing of low-grade ores and materials recycling, any technology offering a means of reducing composite mixtures to their basic constituent components merits serious study.

It appears that a standard fusion torch—that is, a CTR adapted for burnup and recycling of industrial wastes—will prove inferior to alternative recycling techniques. Less costly, more energy-conservative techniques of industrial scavenging seem feasible. However, the promise of the Gough-Eastlund concept becomes more attractive when attention is fixed on a particular category of wastes—toxic residues of nuclear plants. If the neutrons that pack most of the energy from fusion reactions can transmute fertile uranium into fissionable plutonium, they might also be used in more ambitious projects of nuclear transmutation. The objective of what might be called a "radwaste fusion torch" would not be to reduce fission products to recycleable forms—probably an impossible task—but to reduce million-year radwaste disposal problems to 1000-year problems.

From the earliest days of the Atomic Energy Commission's Division of Reactor Development, nuclear power has been promoted as a source of power. Perhaps the development of CTRs should be dominated not by concern with the productive side of the energy equation but by a desire to clean up the accumulated mess of prior nuclear technologies that were developed with an eye to the output of usable electric power.

If radioactive wastes from fission plants could be "bred" into less hazardous substances, a primary initial utilization of fusion technology would be to scavenge some of the radiological debris that will have been created by the nuclear industry through the mid-1990s. In this way, the Gough-Eastlund fusion torch idea might be converted into a concept by which fusion technology could be applied to achieve the goals of the Rose-Kubo radwaste disposal scheme from Chapter 3.

Suppose that all fission reactors henceforward brought into active service as electric producers are configured so that their wastes can be subjected to extremely disciplined control: either reprocessed on site and then recycled back into reactors for burnup (a la Rose-Kubo), or else carefully tended in the immediate vicinity of the plant. Eventually, two or three fusion torches modified to serve as actinide burners rather than as industrial-waste scavengers might be located in power parks specifically designated as clearing points of the nuclear industry. These torches could be used to reduce the longevity of the highest-level wastes from LWRs and from plutonium-producing hybrids. In this way fusion might be used to get the "nuclear house" clean of at least the longest-lived forms of radioactivity. This scenario would assume the decommissioning in the early twenty-first century of fission plants as they begin to wear out. A moratorium might then be imposed on further new construction of fission electric stations as soon as controlled fusion power becomes commercially feasible.

CTRs designed to operate primarily as actinide burners might be frightfully expensive. But the costs of running a device that could transmute much of the collected debris of the fission era into significantly less dangerous sludges—that is, to convert million-year radwastes into thousand-year wastes—could well be sustained as a one-time cleanup effort.

As with a program to develop fusion-fission hybrids, an effort directed at waste scavenging based on fusion torches need not detract from progress toward thermonuclear electric generation. Fusion engineers refer to a "small window" for the achievement of *all* the configurations possible with CTRs. In other words, the conditions required of a fusion device in order to hybridize it with a fission reactor lie on the path to—and not far from—satisfaction of the conditions of plasma temperature, density, and confinement time needed for a fusion electric plant. The same applies to the fusion torch concept. Moreover, the use of a CTR as an actinide burner is not in principle inconsistent with its hybridization or with electric generation. Although a given device might be optimized for one function or another, all functions might simultaneously be performed.

Fusion Electric Plants: The True "Second Sun"

Even though all variants of working CTRs would probably be framed within a single small window, fusion breeders and fusion torches can probably be developed short of solving all the technical challenges of achieving the critical ignition conditions for economical power output. Relatively modest incremental improvements in fusion technology would then be needed for fusion electric plants. But even that extra 10 percent or so could present formidable engineering problems. For this reason, none but the most optimistic plasma physicists predict an ability to produce significant quantities of commercial electricity until the twenty-first century. But the quest seems more than worth the time and the cost.

First among the advantages of fusion is the virtual inexhaustibility of cheap fuel sources–initially, deuterium from water and tritium bred within the CTR itself, using lithium extracted from brines or from the earth's crust. Eventually, it is to be expected that fusion technology could proceed to the deuterium-deuterium reaction, eliminating reliance on lithium as an intervening source of tritium. The D-D reaction would simply obliterate all known constraints on man's energy supply.

What is more, fusion power promises major capital and operating cost savings by comparison with fission electricity. Because vessel ruptures pose less of a threat than is the case with light-water reactors, operating pressures may be increased in a CTR. Hence thermal efficiencies can be higher–up to 60 percent, almost twice the level achievable with LWRs. Containment and shielding requirements are lower in CTRs, making for simpler electric plant systems. And the unprecedented size of a CTR–devices under active study may be five times as big as the largest light-water reactors–should permit design for significant economies of scale.

These factors portend fusion devices built with significantly less investment per kilowatt-hour than will be possible with fission reactors. Table 4-2 shows a modest 10 percent capital saving (from 25.1 to 22.6 mills per kilowatt-hour, mostly accounted for by the reduced shielding and containment requirements of

Table 4-2
Estimated Cost Savings Possible with the Transition from Fission to Fusion

	Mills per Kilowatt-Hour	
	LWR (From Column 2, Table 2-1)	*Controlled Thermonuclear Reactor*
1. Capital Cost of Plant	25.1	22.6
2. Transmission, Distribution	1.1	2.2
3. Fuel Cycle Costs		
Fuel Preparation	2.35	.02
Fuel Recycle	.25	–
Inventory Charge	1.25	–
4. Environmental Costs		
Water Pollution	.5	.4
All Other	.003	–
5. Other External Costs		
Sabotage and Diversion	.053	–
Accidents, Explosions	.002	.001
Occupational Hazards	.01	.001
Low-Level Radiation	.001	.001
6. Transport, Disposal of Wastes		
High-Level Radwastes	1.06	–
Low-Level Wastes	.1	.1
	31.779	25.323

a CTR), plus a major fuel cycle cost saving (from 3.85 to 0.02 mills per kilowatt-hour, to be achieved by switching from reliance on ever leaner uranium ores to reliance on common water as a CTR's basic fuel source). Further economies in the shift from fission to D-T fusion can be registered in such categories as liability to sabotage and radwaste management.

These gains bode to yield an overall saving in the range of seven to eight mills per kilowatt-hour of electricity—a 23 percent saving in the cost of electricity from the replacement of LWRs with CTRs.

The savings imputed to fusion power in Table 4-2 obtain despite an assumed doubling in transmission costs. Although some CTRs may be five times as big as light-water reactors of current design, it seems likely that a typical tokamak-type CTR will come on line at twice the size of an LWR. In this case, fusion reactors will have to be introduced at the rate of a single new tokamak plant for every two LWRs retired. Hence each fusion unit may have to deliver power over approximately double the area served by a typical fission plant.

In almost every category other than transmission, then, fusion power promises to show a significant cost reduction. Indeed, a notably larger reduction can be realized with the shiftover to fusion than may be achieved with the replacement of fossil-fired by fission power. As a consequence, in an energy economy of the sort projected for America in the year 2020—with up to three-fourths of the nation's nuclear capacity supplied by fusion power—the estimates in Table 4-2 suggest a total annual saving to society (measured in mid-1970s cost levels) of some $54 billion.

It perhaps needs to be stressed that the figures in Table 4-2 are only estimates—and estimates, moreover, based in the main on computations by profusion researchers associated with the Plasma Physics Laboratory at Princeton University. Even after discounting for possible bias, however, and after allowing for certain necessarily conjectural assumptions about CTRs that are as yet only designers' blueprints, the relative magnitudes of cost figures strongly suggest that major savings will follow from the switchover to fusion.

Estimates such as the "0.4 mill per kilowatt-hour" charge for water pollution, for example, reflect an optimistic (but reasonable) assumption about the thermal efficiency of fusion reactors relative to fission. On the other hand, the order-of-magnitude gains in such categories as occupational hazards and high-level radwaste management reflect inherent characteristics of fusion—the fact that CTRs imply complete elimination of exposure to radioactivity of the kind represented by a light-water reactor's fission products. In all, then, the figures presented in Table 4-2 seem relative proof against charges that they will probably prove as unrealistically optimistic as were the early estimates that fission-based electricity would be too cheap to meter.

Fusion Reactor Safety

The fuel elements in a CTR (even in most hybrid configurations) never "go critical," and therefore cannot produce a radioactive runaway. Nor does a fusion plant (except in some hybrid configurations) produce bomb-grade material. And since deuterium and tritium cannot be made to explode, the risks of materials diversion by terrorists should diminish as fission gives way to fusion. Yet some radioactive hazards do remain.[8]

In a tokamak-type CTR based on D-T fusion, the maximum accident would occur if the full plant inventory of lithium–some 500 tons of it, including standby stores as well as the quantity used to breed tritium within the reactor core–were suddenly to explode. Like sodium, lithium is highly reactive. Since 0.5 pounds of lithium contains about the potential energy of a barrel of oil, an explosion would have a blockbuster effect, leveling most nearby structures. Released tritium would then oxidize to form a cloud of radioactive "tritiated water," entering groundwaters and polluting the air downwind of the plant.

Even such an event would probably present a less severe public health menace than would an LWR reference accident–not least because the biological half-life of tritium is much briefer than is the hazardous life of released fission products. What is more, the venting of a CTR plant's entire tritium inventory is credible only in the case of purposeful release by a saboteur or lunatic who not only gains access, but who has uninterrupted opportunities to destroy both the tokamak and all standby tanks–without getting himself blown up in the process. If the maximum *credible* accident were to occur–explosive discharge of the tritium in the reacting plasmas and the lithium breeding blanket of a 2000-megawatt tokamak-type plant–the maximum dose estimated for an organism at a 1500-yard fenceline would never exceed 100 millirems. Under normal CTR operations, the maximum annual fenceline dose from airborne tritium releases need not exceed 3 millirems, plus another 0.25 millirem per year to an individual who drank all of his or her water from the dilution pond to which any liquid tritium effluent might be vented.

As the D-T plasmas react within the core of a CTR, the inner wall of the toroidal vacuum chamber, probably made of vanadium- or niobium-steel, becomes irradiated. Neutron bombardment causes the impacted components to become radioactive. The inner wall components also absorb both helium (produced in the plasma reactions) and tritium, further contributing to their embrittlement. The embrittled components, because subject to cracking under stress, must be replaced on a regular schedule. And of course, eventually the entire structural inventory of a CTR will have to be retired.

The wastes from a CTR, whether removed in scheduled component replacements or at the time of final decommissioning, are measurably less toxic than are most fission products from LWRs. Moreover, the irradiated portions of an inner wall would leave the plant in solid form rather than as liquids, as do the fission products from a light-water reactor. This in itself promises to reduce the problem of fusion radwaste control. On-site storage for 10 years can be used to bring the hazard of irradiated vanadium elements to an acceptable level and to reduce that of niobium by several orders of magnitude. In an optimal control plan, irradiated components from a CTR would, following storage to reduce their toxicity, be melted down to release any trapped helium and tritium. The helium is itself a valuable resource, and the tritium could be recovered for recycling as fuel for the D-T reactions. The molten metals might then be refabricated as CTR replacement parts. Such a plan would best be executed in the context of a concentration of reactors with co-located industries employing nuclear-skilled workers—in other words, in the context of a "city of the second sun."

Summary: The Promethean Progression

The progression implicit in the past record of nuclear power points toward human achievement of increasingly Promethean powers—a fundamental change in the conditions of human life, made possible by the harnessing of the power of the second sun. This image of a Promethean progression, tending toward ever more ample supplies of power for human use, emerges as the chief lesson to be drawn from a survey of present and projected nuclear technologies.

Such a progression seems likely, but far from inevitable. Fusion research could still become a stupendously costly disappointment, especially if the first electric-generating CTRs prove to be prohibitively expensive; or if certain materials limits (exotic metals, or ample lithium to breed tritium) cannot for some reason be overcome; or if fusion reactors become so complex as to present a maintenance and operations nightmare. But on the plasma scientists' record to date, optimism justifiably gathers with time.

If successful, the quest for the fires of the second sun will deliver virtually inexhaustible power. Light-water reactors may then appear to have been a mere preliminary form in man's development of nuclear power. The fission era will have represented a learning phase, a period of adjustment preparatory to the spread of fusion electricity as man's primary source of useable energy. As we shall see in Part II, cities of the second sun, in which devices representing the successive generations of reactors could be gathered in complexes containing all elements of nuclear development, would be fully consistent with such a scenario.

Notes

1. "The Alfven Memorandum," 37 *Bulletin of Atomic Scientists* (September 1971), 36.

2. For an account of the timetable for LMFBR and CTR decisionmaking, see the ERDA "National Plan for Energy Research, Development and Demonstration, . . . 1976" (ERDA 76-1, April 15, 1976), pp. 66-67.

3. A brilliant introduction to the exotica of plasma physics will be found in

Samuel Glasstone, *Controlled Nuclear Fusion* (U.S. Atomic Energy Commission, 1974), 88 pp., including illustrations and a bibliography. Almost as lucid is *Fusion Power: An Assessment of Ultimate Potential*, WASH-1239 (U.S. Atomic Energy Commission, February 1973).

4. "Production of Thermonuclear Power by Non-Maxmillian Ions in a Closed Magnetic Field Configuration," 26 *Physical Review Letters* (May 1971), 1156. This work led to Furth's and D.L. Jassby's "Relaxation of the Break-Even Conditions for Fusion Reactors Heated by Reacting Ion Beams," 32 *Physical Review Letters* (April 1974), 1176.

5. Robert L. Hirsch and William Rice, "Fusion Power and the Environment," draft information paper available from the AEC–now ERDA–Division of Controlled Thermonuclear Research, p. 20.

6. See especially Lawrence M. Lidsky's seminal paper "Fission-Fusion Symbiosis: General Considerations and a Specific Example," presented at the 1969 B.N.E.S. Conference at Culham Laboratory, England. In "A Review of Fission-Fusion (Hybrid) Concepts," 20 *Nuclear Technology* (December 1973), 161, B.R. Leonard reviews the Lidsky scheme as well as other proposals for rapid development of a working hybrid system. Also helpful are E.L. Dryer and S.J. Gage, "The Fusion-Fission Breeder: Its Potential," AEC Symposium on the Technology of Controlled Thermonuclear Fusion Experiments (Texas, 1972); R.W. Moir, "Mirror Fusion-Fusion Reactor Designs" (Lawrence Livermore Laboratory, UCRL-78629 Reprint, August 1, 1976): F.H. Tenney, "A Brief Review of the Fusion-Fusion Hybrid Reactor" (Princeton Plasma Physics Laboratory, unpubl., October 1, 1976).

7. "The Prospects of Fusion Power," *Scientific American* (February 1971). Lidsky develops the idea in his conceptual study of an "Augean Reactor." See his "Fission-Fusion Systems Hybrid, Symbiotic and Augean," 15 *Nuclear Fusion* (February 1975), 151, pp. 168-172.

8. On the environmental impacts of a CTR, see especially S.E. Beall, et al., *An Assessment of the Environmental Impact of Alternative Energy Sources* (Oak Ridge National Laboratory, ORNL-5024, September 1974), at pp. 102-103.

Part II
Getting from Here to There

Introduction to Part II

A dominant theme of Part I has been that nuclear policy should be considered in light of the eventual promise of fusion power. Meanwhile the fission era presents certain grave issues that challenge the wisdom of those who will preside over the transition to fusion. Imprudence or improvidence in the management of fission technology could, between now and the turn of the century, so pollute and impoverish the globe as to make a fraud of the promise of atomic power.

The task of contemporary nuclear policy is to maintain the Promethean progression while minimizing the peculiar hazards of a transitional fission era. The problem will be to get from here to there–from fission to fusion–without delivering the instruments of global holocaust into the hands of irresponsible leaders, without poisoning the environment with radioactivity, and without making posterity hostage to the fallibilities of a "nuclear priesthood,"[1] whose members would ritually tend the long-lived waste products of fission power. We must now consider how power parks can be used to approach this dominating problem.

The fission era will probably extend over the years 1980 to 2040, with its zenith just about at the turn of the century. Power parks could become the governing institution of the years of transition from fission to fusion. As such, power parks could function as the keystone of America's domestic energy policy between now and the year 2000.

We have seen that the case against the atom depends heavily on the validity of charges that nuclear power is both too expensive and too dangerous. Any policy which satisfies the criterion of comprehensiveness must respond to these charges. We shall see in Chapter 5, on the basis of adjustments in the cost estimates already presented (in Tables 2-1, 4-1, and 4-2), that the clustering of reactors in power parks would promise modest savings in the fission era–savings that the transition to fusion could multiply. What is more, the potential of power parks to contribute to the safeguarding of nuclear materials was cited by early proponents of the concept as an even stronger argument than that of potential cost savings. The issues raised in Chapters 2 and 3, then, find partial resolution in the policy discussed in Part II. And so does the issue raised in Chapter 4, the issue of the transition to fusion.

A program to maximize prospects for CTR-based electricity does not depend in any direct way on the adoption of power parks. But the development of CTRs, as well as chances for a smooth transition to commercial fusion power, *do* depend in considerable measure on the maintenance of a vigorous nuclear industry in the 1980s and 1990s. An active nuclear industry implies a continuation of strong academic programs in nuclear and nuclear-related fields–the requisite of a trained manpower base and of a broad-scale program in basic research. It implies a buildup of experience in nuclear engineering and in solving various problems that are common to fission and fusion plants but which do not occur in other forms of electric generation. Such problems include the handling of irradiated materials, the engineering of unprecedentedly capital-intensive thermodynamic systems, and perhaps the complex issues of breeding. As a direct contribution to America's fission program, then, the power park concept could indirectly contribute to the installation of a fusion economy as well.

America's fusion plants will have to be introduced into carefully prepared environments. They will need highly skilled construction and operating crews. They will need ample land areas and cooling water in abundance. (Or what may prove an even more formidable challenge, fusion electric plants may need costly nonevaporative cooling devices—"dry towers"—in the event that earlier generation electric stations preempt the available cooling water between now and the year 2000.) Unless technological breakthroughs permit the transport of electricity by microwave beams relayed from orbiting satellites, the fusion plants of the early twenty-first century will need new transmission corridors and high-voltage lines. And they will need a national logistical net to serve demanding material support requirements. Provision for these needs is implicit in the technical and social planning that will be required for power parks. Thus the fission-based parks likely to be under construction by the early 1980s appear as logical sites for the introduction of the fusion plants of the early twenty-first century.

Power parks, even while serving as templates for the pattern of a fusion economy, may become base industries for new towns and for the economic stimulation of existing communities. These communities, which will initially draw their economic support from nearby power parks, merit the title, *cities of the second sun.*[2] As we shall see, some 27 parks might be built. Each power park could, if sited in rural America and purposely used as a growth pole for regional economic development, account for the eventual in-migration to its immediate vicinity of up to a quarter of a million people. Demographic data suggest that population shifts of this magnitude could help reverse the decline of selected regions in midcontinental America.

This, then, is the thesis: America's need for energy can best be approached in the context of an intelligent program of domestic civilian atomic development. In turn, the full potential of nuclear energy in America can best be realized—and the hazards discussed in Part I best be minimized—in the context of planning for power parks as an alternative to the familiar pattern of small dispersed nuclear electric power stations. The power park concept can be used to harmonize an entire set of innovative approaches to nuclear siting, nuclear safety, and nuclear cost control. And the full potential of power parks can be realized only in the context of planning for cities of the second sun.

Notes

1. The expression is Alvin Weinberg's. See his "The Moral Imperatives of Nuclear Energy," 14 *Nuclear News* (December 1971), 33.

2. For an overview of the argument presented in Chapters 6-7, see my "Community Implications of Energy Parks: A Regional Development Perspective," General Electric's *Assessment of Energy Parks vs. Dispersed Electric Power Generating Facilities*, Vol. II (National Science Foundation, May 30, 1975), pp. 6-104 through 6-176.

5 Power Parks as Institutions of the Transition

While no economic returns can be anticipated from the transition to fusion until the early twenty-first century, an improved reactor siting strategy during the transitional era of fission power could produce significant savings as early as the late 1980s. The concept of power parks[1]—whereby multiple reactors along with supporting fuel cycle facilities would be clustered on integrated sites far from existing urban centers—offers the best chance to effect further savings. An optimum strategy for the transition, then, translates into a national policy favoring reactor development not in isolated dispersed sites selected by the electric utilities on the basis of their own operating convenience but in carefully planned nuclear centers.

The nuclear center concept may be regarded as a logical outgrowth of trends in the industry since the early 1960s. A mass of fissionable material that just exceeds the size needed for criticality can produce enough heat to generate some 200 megawatts of electricity. Most early American reactor designs looked to devices of this "minimally critical" scale. However, in the early 1960s researchers at Oak Ridge National Laboratory documented the economic gains that might be achieved by putting several critical masses into a single unit. Since then, planners have stressed the advantages of plants at ever larger, more efficient scales. The "reach for scale" promises to continue through the progression from fission to fusion. As noted in Chapter 4, the biggest contemplated thermonuclear electric stations may be five times as large as are the costliest fission plants now on line, and these exceed the nuclear stations of the early sixties by another factor of five. Nor is any end in sight.

Before the year 2000, the reach for scale will probably culminate in large clusters of nuclear reactors. These concentrations will be based on updated versions of plans developed in the late 1960s within the U.S. Atomic Energy Commission for major aggregations of nuclear generating capacity in single locations. The AEC studies purported to show (perhaps, skeptics charged, they were *designed* to show) that American nuclear technology had significant potential as an export product.

Nuclear proponents envisioned agroindustrial complexes as huge factory-and-farming developments on the banks of the Red Sea, the Australian coast, the River Ganges. Atomic plants would desalinate water and provide electricity for fertilizer synthesis to make deserts bloom or to feed India's masses. In 1970, a survey of agroindustrial complex studies released by Oak Ridge National Laboratory promised a 68 percent return on investment in a development designed for India that would include an ammonia plant, a phosphorus plant, and an aluminum plant, with enough electricity left over to power 25,000 tube wells for local irrigation.[2]

A decade after the first agroindustrial complex studies were begun, anticipated shortfalls of electricity in the United States directed interest again to the potential of power parks (also called *nuclear centers, energy parks,* or, as we shall see, even *solar farms*). But the market now lay not across the seas. Power parks were seen as potential solutions to the problems of domestic American energy policy. Allegedly, these complexes would achieve economies of scale in plant construction, plus significant efficiencies in the operation of central electric stations.

This chapter considers power parks as dominant institutions in the electric power industry during the transition from fission to fusion.

Nuclear Centers: The Politics of an Idea

As early as 1943, the imaginative and politically astute physicist Alvin Weinberg, with his mentor the Nobel laureate Eugene Wigner, had drawn up the first plans for a breeder reactor. Weinberg eventually reached one of the most prestigious positions in the American scientific community, the directorship of the Atomic Energy Commission's Oak Ridge Laboratory.[3] Weinberg's faith in the breeder never waned. Successful development of the breeder, he argued, held the secret of virtually limitless cheap energy for all mankind. It was as a solution to the problem of devising a safe siting configuration for breeder reactors that the "power park" concept emerged.

Given a deepening public suspicion of the candor and even the competence of nuclear industry leaders, proposals for new technological extravaganzas—such as a full-scale commitment to the breeder—would face extremely rough review, if indeed they could be fairly considered on their merits at all. The proponents of novel schemes could no longer rely on the uninhibited funding and unchallenged political support, especially in Congress, that had characterized the atomic enterprise in earlier years. Continued reactor development, environmentalists charged, would mean unchecked economic growth. It would elevate technocrats in the nuclear lobby to dangerous new heights of power. It would—and in a country now committed by the National Environmental Protection Act of 1968 to preserve and enhance the natural environment—require the condemnation of more and more acreage for cooling towers and radioactive-waste disposal sites.

Weinberg, seeking to turn the gathering momentum of antinuclear forces to the advantage of the breeder, adopted the AEC's old agroindustrial complex idea. But he adopted it with a new twist, one calculated to confer the quality of credibility on the power park concept.

Deemphasizing the original industrial and agricultural implications of nuclear centers, Weinberg stressed the contribution to safety and safeguards that might be achieved by co-locating all reactors (including breeders as they come into service) along with fuel reprocessing facilities:

> The point here is that if, say, 10 reactors are sited close by, then the entire installation, by virtue of its size, might be expected to possess technical and engineering resources that would be impractical in a small installation. Should a serious accident occur in one of the reactors, the resources of the entire installation could be mobilized to deal with the incident, and to confine the spread of radioactivity. I realize that such nuclear parks may be more vulnerable to common-mode failure such as enemy action or earthquake; nevertheless, the balance seems to me to lie on the side of easily mobilized resources. This, at any rate, is the impression I get from the experience at Hanford and Savannah River: nuclear parks which are able to mobilize with impressive swiftness and efficiency.[4]

Weinberg further contended that the most serious safeguards problem might be solved by gathering all the nuclear facilities of a region into a single complex. As suggested in the diagram of Figure 5-1, plutonium from breeders could be recycled into reactors also sited inside the same fenceline.

To a considerable extent as a consequence of Weinberg's influential advocacy, the concept of nuclear centers gained support among federal officials. Under a grant from the National Science Foundation, the Center for Energy Systems of the General Electric Corporation conducted a 9-month study of the power park concept.[5] General Electric, a prime supplier of boiling-water reactors both to American utilities and foreign purchasers, could hardly have been regarded as altogether disinterested. But the G.E. study team avoided consideration of the desirability of nuclear growth. At issue was the economic desirability of parks as compared with dispersed siting both of fossil-fired and nuclear plants. As was underscored in Chapter 2, the problem of rising costs has hurt nuclear suppliers by dampening demand, rather than helped them by widening profit margins. Thus the economic incentives applicable to a firm such as General Electric worked in the direction of hard-headed, fair analysis–if not of the virtues of growth in the nuclear industry, then at least of power parks as a way to put the industry back on a growth path.

The G.E. evaluation, published in May of 1975, became the basis for a series of conferences on power parks sponsored by the Energy Facilities Siting Panel of the National Science Foundation, whose Director simultaneously served as the principal advisor to the President on science and technology.

The prototype nuclear center considered in the G.E. study would provide for 26,000 megawatts of generating capacity at the conclusion of a 17-year construction schedule. The park shown in Figure 5-2 could contain BWRs plus a few LMFBRs or hybrids, with the breeders phased in on a schedule to ensure an equilibrium balance with the BWRs. That is, with adequate co-located reproces-

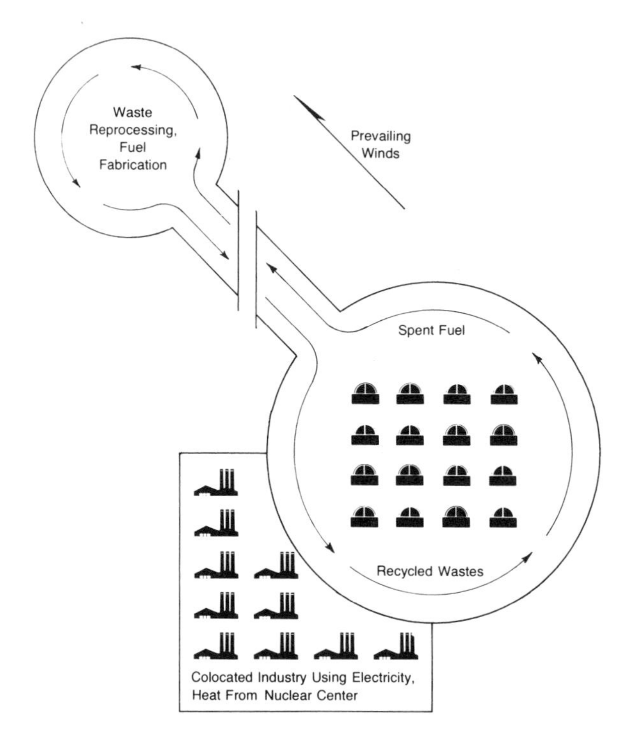

Figure 5-1. The Closed Reactor-Reprocessing Fuel Cycle

Notes to Figure 5-1. General Electric Company researchers have estimated that 20 reactors, each of approximately 1,000 Mwe capacity, could produce enough radioactive wastes to support a co-located reprocessing plant. Such an operation would be at the lower limit of economic feasibility. More reactors (or, of course, the same number of reactors, but with each of greater rated capacity) would make the operation more attractive financially.

The addition of a reprocessing plant and a fuel fabrication facility would necessitate the dedication of an additional 100 to 115 acre exclusion area. The most hazardous nuclides–plutonium and other actinides–would then never leave the perimeter of the park, since these would be subject to constant recycling.

Reactors within the nuclear generating module of the power park would probably be arranged in quads of four units apiece. According to U.S. Nuclear Regulatory Commission data, spacing of these quads approximately two and one-half miles from one another should reduce both the effects of the park's heat load and of its routine radiation releases to levels no higher than would occur if an equal number of quads were randomly dispersed.

The maximum radiological exposure of an organism permanently stationed at the fenceline would fall well below the 100-millirem level of exposure typically experienced by all individuals from natural sources. Adoption of the dumbbell configuration, together with thicker building shielding, could further reduce the radiological load on the immediate neighborhood of the complex.

In the dumbbell model, plutonium reprocessing and fuel fabrication facilities are removed along the vector of prevailing winds in the direction opposite from that of any planned industrial and community development. The connecting artery, protected by alarms and fences from trespassers, is used only by special trucks bearing spent fuel from the generating station, and carrying actinide-packed recycled fuel rods back to the reactors in return. Infrequent trips would be made on irregular schedules to help ensure against would-be hijackers of nuclear materials. This dedicated roadway would have no intersections, but perhaps an overpass or two.

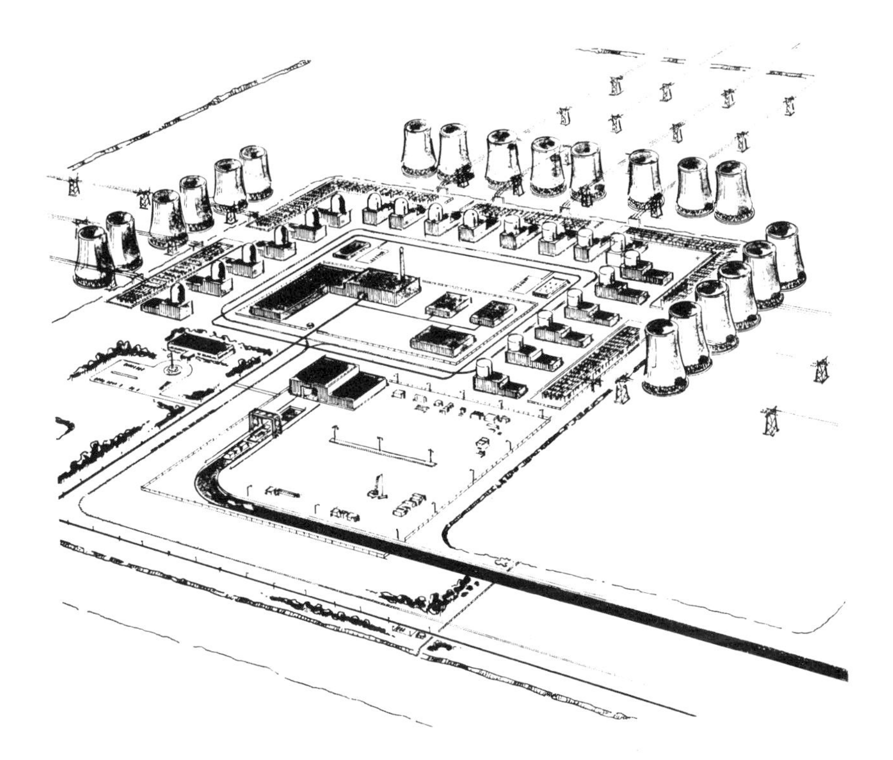

Source: General Electric Corporation, "N.S.F. Energy Park Study," 5-30-75. Reprinted by permission.

Figure 5-2. Artist's Conception of a Nuclear Center.

Notes to Figure 5-2. In this artist's conception of a 20-unit nuclear center, reactors are concentrated on a manmade island surrounded by cooling water. Wet cooling towers rise from an artificial pond. These towers provide primary cooling for the park, reducing the area required for the pond. "Get-away lines" bearing off high-voltage electricity have twice the capacity required by the park. Thus even if half the lines were short-circuited, the full electrical output of the 20 units would be delivered to market.

In their Nuclear Energy Center Site Survey mandated by Congress in the Energy Reorganization Act of 1974, members of the Nuclear Regulatory Commission staff presented a "conceptualized description" of the power park concept as it might apply to a larger nuclear complex than the one shown. Figures describing the physical dimensions of such a center, as well as the likely environmental impacts and financial resources required, hint at the impact of a national policy favoring nuclear centers.

The estimated capital costs of the biggest envisioned facility (48,000 Mwe) would range from $30 billion to $45 billion. Once built, the facility would imply an annual operating budget of $2 billion, give or take a half-billion.

Up to 80 square miles would be required for the reactors and generators, switch yards, a radioactive exclusion area, cooling equipment, and fuel recycling facilities. An additional 200 square miles might have to be dedicated for use as transmission corridors. Nor do these figures indicate the acreage needed to house the construction crew (which would reach 11,000 workers at peak), or any new community development which might be stimulated by the siting of the park.

In the NRC's "conceptualized description," reactors would be clustered in groups of four, with these quads separated by several miles from one another in order to facilitate heat dissipation and to ensure adequate local dilution of any radioactive effluents.

sing and fuel fabrication facilities, the amount of bred plutonium would just equal the required replacement fuel loadings of the light-water reactors.

Power parks also admit of physical configurations other than concentrated reactor siting in single locations. Figure 5-3 illustrates two alternative configurations, either of which would achieve certain of the economic and safety-related gains of concentrated nuclear centers. Yet these configurations—"satellite" and "lattice" siting—would avoid some of the disadvantages that have been alleged against the original concept of nuclear centers, such as the disadvantage of having to secure purchase rights covering a single huge parcel of land (up to 80 square miles) in a given region.

In order for a new concept, such as that of power parks, to advance beyond the level of a mere glint in some engineer's eye, it must gain chaperonage through that most hostile and change-resistant of environments, the Washington bureaucracy. Normally after a preliminary feasibility study—in the case of power parks, the National Science Foundation study conducted by General Electric Corporation researchers—a concept that has gained basic support at high political levels gets assigned for "staffing" by a responsible agency. But in a bureaucratic environment marked by a crazy quilt of agencies with overlapping interests in energy problems, which was the "responsible agency"?

Ostensibly, the Atomic Energy Commission was the responsible agency. But the AEC was itself falling victim to fragmentation. By the Energy Reorganization Act of 1974, the regulatory or "watchdog" functions of the old AEC—focused especially on reactor safety and safeguards—were transferred to a new federal agency, the U.S. Nuclear Regulatory Commission.[6]

Nuclear promotional activities—which, critics of the AEC charged, had dominated that organization's activities to the detriment of reactor safety—were incorporated in a newly created U.S. Energy Research and Development Agency. ERDA also gained promotional and developmental responsibilities in other energy areas, including fossil fuels and solar power. So nuclear fission would have to compete on its merits with alternative fuels. Regulatory and promotional activities now confronted each other in a kind of adversary deployment of the nation's bureaucratic resources within the NRC and ERDA.

To ensure credibility in further studies of the nuclear center concept, pronuclear congressmen asked that the next follow-on action be delegated to the Nuclear Regulatory Commission rather than to the promotionally oriented Energy Research and Development Agency. The 1974 Reorganization Act mandated a national survey of potential locations for nuclear centers to be conducted along with an evaluation of power park economics by the new NRC's Office of Special Studies.

The Economic Case for Nuclear Centers

On January 19, 1976 the five commissioners of the NRC sent Congress the "Nuclear Energy Center Site Survey."[7] The study included analyses of a wide

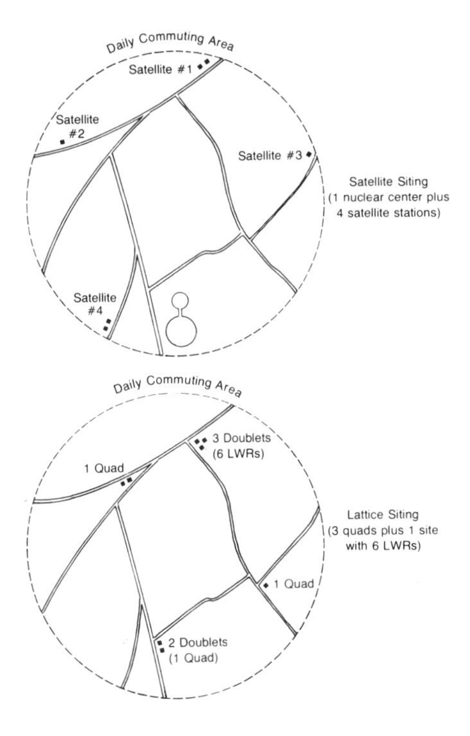

Figure 5-3. Satellite and Lattice Siting of Nuclear Plants.

Notes to Figure 5-3. In aggregative siting–the conventional image of a nuclear center–all new generating capacity in a given region would be concentrated at a single site. At the opposite extreme, dispersed siting refers to the grouping of up to four reactors on each of a series of separate sites, without regard to any goal of maintaining a specific national or regional electric development pattern.

The illustration depicts two modes which lie between these extremes. With *satellite siting,* a "primary" nuclear center (probably of from 6000 to 20,000 megawatts) would be developed at a suitable location. Over a period of years, even decades, the construction force would then build a series of satellite nuclear stations–all within daily commuting distance of a common set of communities near the primary center. The generating capacity ultimately brought on line within the commuting area–indicated by the dashed perimeter line–might equal the total capacity which, with aggregative siting, would be concentrated at a single nuclear center.

In *lattice siting,* quads or doublets of reactors would be carefully developed within the boundaries of a single commuting area, spatially arranged to take maximum advantage of the regional transportation system so transfers of personnel and materials among the "miniparks" would be facilitated. Assuming an ability to preserve significant elements of design standardization–and hence, learning curve effects within the regional workforce–it should be possible to recover with lattice siting most of the manpower associated benefits of the power park concept.

If the electric capacity within a region were to be developed in either the satellite or the lattice modes, rather than in the aggregative mode, some additional expense would probably be required to develop the regional transportation network. Because reactor construction requires more highly skilled operatives–and operatives who are, moreover, trained in rather different trades than are needed in road-building, e.g., pipe fitters and welders qualified to meet nuclear safety standards–the workforce needed to build an efficient regional commuting system would probably be additive to that required for construction of the power park itself.

range of possible power park configurations, from the siting of reactors in quads (four 1,200-Mwe LWRs at single locations) up through monstrous aggregations (40 nuclear units, together with full plutonium reprocessing and fuel fabrication facilities, all sited within a single fence).

The NRC analysts found that in most circumstances, nuclear centers consisting of from 15 to 20 reactors more than "paid their way." The relative merits of parks at megawattage levels much bigger or smaller than 15 to 20 units varies from site to site.[8]

Efficiencies of two sorts may be expected. First, reorganization of the generating and transmission segments of the electric industry might permit more efficient operations along regional lines.[9] Second, capital savings from more efficient on-site construction procedures may more than compensate for the higher transmission costs implicit in the concentration of generating capacity.

The Regionalization of Electric Planning

Electric power utilities are in the main "vertically integrated." Typically, a single company will provide not only for electric generation and then long-distance transmission to the load center but also for final distribution of electricity to the consumers.

For years critics have charged that the electric industry evidences neither the professionalism nor the sense of public responsibility required in a future likely to be marked by increasing dependence on nuclear power. A decision for nuclear centers would lend impetus to trends that already point toward increased reliance on regionalism as an approach to energy planning. A move to power parks—each serving many hundreds of square miles and several big cities—would furnish a natural opportunity to break power generation and transmission away from the function of local electric distribution. The "G-T" phases could then be organized under a newly constituted private management structure in the region, or perhaps under a regional public authority similar to the Tennessee Valley Authority.

If the G-T functions were progressively assumed by new private or public regional electric companies, the existing investor-owned utilities could move toward an emphasis on local marketing and distribution. The G-Ts would generate and transmit electricity from power parks for resale to existing utilities. The utilities would then market the electricity at retail.

At first, a new G-T might plan, construct, and manage only a single power park, leaving undisturbed the ownership of any existing dispersed generating sites within the region. But since all new units brought on line within the region would be built and operated by the appropriate G-T, the overall structure of the electric industry would gradually change as old dispersed generating sites are phased out of service through the normal processes of attrition. The role of existing utilities in electric generation and long-haul transmission would diminish; the regional companies would gain a larger and larger share of control. These regional companies (or "regional electric authorities," if patterned on the TVA model) might most efficiently plan the elaboration of the electric grid.

At present, utilities prefer to site generating facilities according to criteria that are tailored to the interests of their own investors, their own customers, and their own state public utility regulatory commissions. Such criteria, however, can lead to serious inefficiency and waste, as when the advantages of larger-scale electric production are foregone because each utility in a region builds its own dispersed generating stations rather than cooperates with other companies to build more efficient central stations serving all at reduced unit costs. Moreover, the environmental and socioeconomic effects of plant construction may reach beyond the service perimeter or the state line of a given utility. A regional authority could bring the full schedule of trade-offs into perspective when planning new facilities rather than merely those which bear on a single utility's profit-and-loss statements. By no means of least importance, the more spacious view afforded by regional organization would permit comprehensive planning to balance generation and transmission costs—a critical requirement, for it is this balance which decides the economic case for power parks.

Cutting the Capital Costs of Power

A solution to problem number 1 in the American nuclear industry—the challenge of rising costs of fission power—presupposes a reversal of the cost trend of the sixties, years that saw constant experimentation with reactor designs. Design novelties and progressive increases in the sizes of new units prevented plant engineers from realizing either any appreciable efficiencies of standardization or any notable gains from a maximum learning curve effect.

A power park, however, would offer a local "reactor market" large enough to justify the construction of a special nuclear factory located within the region or even immediately adjacent to the generating complex itself. Once the site of a nuclear center has been cleared, standardized techniques may be used for accelerated construction of prefabricated reactor components, followed by on-site modular assembly of units. The General Electric prototype power park—the nuclear center depicted in Figure 5-2—would cost an estimated $31.5 billion in 1974 dollars, as against $38.3 billion if its 20 units were dispersed over a series of two-, three-, and four-unit generating sites.[10] In the G.E. analysis, a saving of some 20 percent of the construction costs likely if equivalent megawattage were built in the dispersed mode owes to the benefits of modular reactor construction followed by on-site component assembly.

The actual figures are striking: up to 50 percent of the on-site labor may be moved into the special reactor factory, where more efficient standardized production techniques might yield as much as a 30 percent reduction in the number of man-hours needed to build the park in question. A special gantry crane of a thousand tons' capacity would transport modules (or, in the term used by the G.E. analysts, supermodules) from the reactor factory to the actual unit site for installation.[11] As we shall see in Chapter 6, the concept of a regional reactor-module factory suggests important opportunities for using nuclear centers as growth poles to concentrate economic activity in selected areas of the intermetropolitan frontier.

The Nuclear Regulatory Commission study suggested a somewhat more modest level of on-site savings than did the General Electric evaluation. But still, of the 10 to 15 percent cost reduction reflected in Table 5-1, a significant fraction—at least half the estimated on-site savings—results from the more favorable manpower arrangements that should be possible with power parks.[12]

Rather than relying on a gypsy manpower pool, a permanent regional work force could be assembled. Workers could develop the special skills needed for nuclear construction, especially in critical crafts such as welding and pipefitting. It would be reasonable to look for significant learning curve effects, as workers who repeat a given procedure devise improved production techniques.[13] (The General Electric researchers estimated that modularized production combined with learning curve effects would shorten the construction period of an average generating unit by about 17 percent—from 66 months in the dispersed mode to

Table 5-1
Cost Savings Achievable with Nuclear Power Parks

	Percent of Direct Costs Saved by Siting Mode		
	Aggregative (24-Unit Park)	*Satellite 12 + 2 + 4*[a]	*Lattice 6-Quads*
1. On-Site Construction Savings			
Centralization of Support			
a. Special Construction Equipment	3.9	3.8	–
b. Bulk Material Purchase	0.5	0.3	–
Standardization, Modularization	2.2	2.3	2.2
2. Improved Manpower, Contracting			
Reduced Labor Need	3.0	3.0	3.0
Productivity Gains from:			
a. Improved Job Conditions	1.0	1.0	1.0
b. Optimum Skill-Level Mix	2.1	2.1	2.1
c. Stability of Employment	1.0	1.0	1.0
Long-Term Equipment Contracts	1.2	1.2	1.2
	14.9%	14.7%	10.5%

[a]Estimates prorated to reflect a 12-unit primary nuclear center in the region, followed by the construction of 2 quads and 4 single units on so-called handkerchief sites for a total of 24 reactors in the region.

55 in parks.) And added to the directly measurable benefits of a stable labor cadre would be certain intangibles, e.g., the development of esprit in a workforce whose members would identify as residents (not mere builders) of a rising city of the second sun.

The Costs of Long-Haul Transmission

If all the nuclear capacity expected in the year 2000 were met by generating stations consisting half of quads and half of doublets, a national requirement for more than 200 power station sites would result. The grid in the upper portion of Figure 5-4 suggests the density in a national electric network built to support this requirement. Meeting two-thirds of the same projected nuclear demand by reactors clustered in power parks implies a need for fewer than 30 points of intersection.

Shifting from a fine- to a coarse-gauged national electric grid would increase the average distance of a generating center from the loads served by its units. For the gains of geographically concentrated electric capacity, the added distances would exact a price. Remote siting of power parks can translate into a requirement for thousands of acres in transmission corridors.

Land for these corridors would have to be secured either by purchase before ground were broken or else through successive acts of condemnation on a gradual schedule as the park itself develops. In the case of purchase before ground-breaking, decades might intervene between the initial capital commitment and the achievement of break-even in the output of revenue-bearing electricity. During these intervening years, an enormous investment would lie at risk. Should loads fail to develop as predicted, additional costs might have to be sustained not merely to reroute transmission lines but also to buy up more land. On the other hand, with gradual acquisition of transmission corridors by condemnation as loads develop, delays could be encountered in gaining the title to needed land, probably at a considerable ultimate cost penalty.

The best achievable compromise would doubtless require the development of a comprehensive regional plan—a subject to be discussed in some detail in Chapter 7. Though such a plan would itself be subject to continuous revision to meet changing conditions in the region, it would specify likely lands for *future* condemnation, not only to provide transmission corridors but also to provide sites for satellite "miniparks" and even for any industrial developments that regional authorities might seek to promote. Options to condemn at specified future dates could then be purchased before construction began on the original power park. Such a procedure would put landowners on notice regarding the possibilities for future development yet would leave the regional planners as well as industry officials considerable flexibility in determining the course of construction as the economic picture changes.

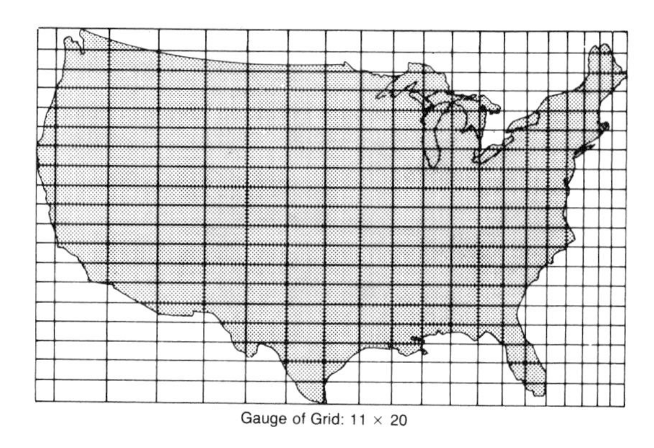

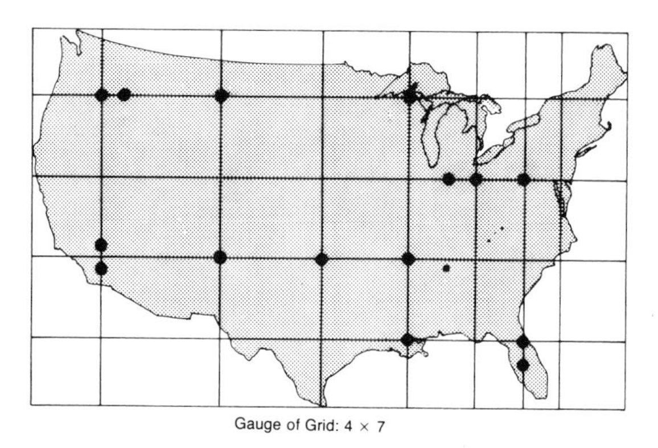

Figure 5-4. Coarsening the Gauge of the National Grid.

Notes to **Figure 5-4.** The upper illustration suggests the fineness of the gauge of a national electric power grid which meets two-thirds of a requirement for 800 megawatts of nuclear electric capacity half-and-half with quads (four generating units per site) and doublets (two per site). The fine-gauged mesh is achieved by crossing 20 vertical bars with 11 horizontal.

With a series of energy parks of about fifteen 1,200 Mwe reactors per park, the number of vertical bars reduces to seven, and the horizontal to four. The net effect is to coarsen the gauge of the national nuclear electric grid by almost an order of magnitude.

Needless to say, the actual siting of power parks would require more than merely overlaying of a U.S. map with a template of the appropriate gauge. However, such an overlay could be used as a point of departure in power park site planning. Actual sites for nuclear centers could then be chosen by distorting the national grid to move specific nodes or intersections out of seismically unstable areas toward available sources of cooling water and away from major population centers, and so forth.

The 15 starred nodes in the lower illustration correspond to locations under active study in 1975 as potential sites of either coal-fired or nuclear power parks. In the East, these sites include several alternatives under consideration for a single (mixed coal-fired and nuclear) park in Pennsylvania, two sites in Florida, and two in Michigan.

In the midcontinental strip, potential sites include the Aitken County coal-fired park in Minnesota, Camp Gruber near Tulsa, Oklahoma, the Tom Bigbee-Pickwick Reservoir to serve markets in Alabama, Mississippi and Tennessee, and one location for a major all-nuclear center north of Baton Rouge, Louisiana.

Washington state and California alike have two locations under study. The remaining sites are at Glasgow Air Force Base in Montana and in the Tularosa Basin of New Mexico.

Twenty-five acres are typically needed per linear mile of long-haul transmission line to bring power from a reactor to the vicinity of the ultimate consumer—almost a square mile of land per 25 miles between the park and a major power dropoff. So land selling at, say, $300 per acre at the time of purchase would cost three-quarters of a million dollars for a single corridor 100 miles long. If a power park served four major loads, each 100 miles away on a different azimuth, the cost would quadruple.[14] So barring future breakthroughs in transmission technology that may reduce the need for land rights-of-way (e.g., low-cost beaming of electric power long distances by laser or microwave), proponents of nuclear centers must be prepared to qualify their claims of on-site capital savings with offsets for remotely sited energy parks.

In sparsely settled regions, where loads are relatively small and distances large, the nuclear center concept must probably be compromised in favor of more smaller stations, each located closer to its respective load center. In western United States, transmission penalties could range from almost $200 million (for 10-unit parks) to four times that (for 40-unit nuclear centers). Of course, even these estimates from the NRC nuclear center site survey report do not necessarily preclude other benefits of power parks—for example, improved safeguards for nuclear materials by closing the fuel cycle, or as economic stimuli in depressed rural regions—from tipping the balance toward nuclear centers even at some net economic penalty.

Transmission penalties much less severe are possible when power parks can be sited closer to major loads, as would be possible in the East and through much of the central United States.[15] According to NRC figures, a break-even point—at which the transmission penalty just cancels on-site cost savings—occurs at about 125 miles from the main load center for a 20-unit power park. At greater distances, the case for power parks may have to be based not on potential capital savings but on the contribution of nuclear centers to noneconomic policy objectives, such as increased safety.

Closing the Nuclear Fuel Cycle

Power parks offer two kinds of gain in the field of radiation safety. Most obviously, by "closing the nuclear fuel cycle," power parks that include reprocessing and fuel fabrication facilities within their fencelines could improve the controllability of high-level radwastes and plutonium. Second, it is possible that nuclear centers could in some instances contribute to the execution of a more responsible plan for radwaste cleanup and disposal.

Co-Located Reprocessing Facilities

Eight hundred thousand megawatts of installed nuclear capacity in the year 2000 would imply some 10 shipments of irradiated wastes casks per day to reprocessing plants somewhere in the United States. The General Electric Corporation study of energy parks suggested a need for as many as 20 reprocessing facilities at the turn of the century.[16] By the mid-1970s, only three such plants had been built, and two of these proved to be failures, owing to technical difficulties in the layouts of plant operations. Even with perfectly safe and efficient plants in place, the reprocessing operation raises grave safeguards issues—issues that gain in urgency with progress toward the second generation of nuclear plants, which rely for fuel on recycled plutonium from LWRs and perhaps eventually on the output of breeder reactors. Shipments of bomb-grade materials all over the country to reprocessing plants and then back to reactors could literally invite diversion by saboteurs or terrorists.

(Adding to the consternation of antinuclear spokesmen are prospects for the eventual introduction of high-temperature gas-cooled reactors as competitors of the LWR. Hopes for rapid commercialization of the HTGR became a temporary casualty of the deteriorating capital-cost situation of the early 1970s. But because of its high thermal efficiency, the gas-cooled reactor remains a most promising contender and perhaps a partial alternative to the breeder. But the HTGR requires highly enriched uranium as fuel—almost 95 percent U-235, as against the 3 percent used in light-water reactors. Such fuel could be made into a bomb. Furthermore, the HTGR converts fertile thorium into U-233, another potential feedstock to the nuclear explosive industry.)

The threat of diversion presents the most compelling technical argument for closing the fuel cycle by co-locating reactors and reprocessing facilities, thereby minimizing the number of access points vulnerable to would-be nuclear thieves.

A reprocessing plant accepts spent fuel rods from reactors. The spent rods are then chopped into firecracker-sized bits to facilitate the extraction of plutonium, unburned uranium, and perhaps cesium and strontium for special disposal. The chopped-up fuel elements are dissolved in acid, releasing radionuclides which then enter either the liquid or the gaseous reprocessing streams. It is these streams that are finally treated to release essentially all remaining krypton and gaseous tritium and to recover other elements for storage or reuse.

A co-located plutonium reprocessing and fuel fabrication facility might have an annual throughput of 1500 tons of spent fuel, to be reprocessed so elements from the treatment streams may be fabricated into mixed oxide fuel elements for recycling within the complex. Such a throughput implies a storage at steady-state of approximately 17 billion curies of radioactivity on-site—that is, in the reprocessing module of a dumbbell-shaped nuclear complex. This level of radioactivity would increase by half if the throughput were derived entirely from LMFBR wastes.

Partly because of the huge radioactive inventory in the fuel cycle, reprocessing plants would also routinely emit significant levels of radioactivity. The one reprocessing plant on which reasonably valid data are available—the facility at Barnwell, South Carolina—should emit no liquid radwastes.[17] But the higher gaseous emissions from reprocessing plants (compared with reactor emission levels) necessitate the condemnation of greater exclusionary acreage around a fuel cycle facility than might be needed around a park consisting only of reactors.

The Barnwell "exclusionary zone boundary" is some 2400 yards at its shortest radial distance from the fenceline to a building within the facility. With a 1500-ton throughput, Barnwell will routinely emit krypton, tritium, fission products, and even trace quantities of the actinides (including five-thousandths of a curie per year of plutonium). With the sole exception of some iodine releases, which imply approximately 150 man-rems of exposure annually in the general region of the facility, the Barnwell emission rates suggest exposures well below the lowest permissible levels ever promulgated by public health authorities, including officials of the Federal Radiation Council, whose members are not regarded as captives of the atomic energy industry.

Reprocessing plant engineers—who seem also to have been immunized against technological pessimism, although their record to date is signally less impressive than is the plasma physicists' record of experimental progress—look to a tenfold improvement in iodine removal systems in order to bring release levels of this nuclide into line with those of actinides and other fission products.[18]

All the main routine emissions of a reprocessing plant enter the body through the food chain. Hence the exposure of organisms in the area could be reduced by various techniques aimed at interrupting the penetration of the local food web.

These techniques might include the condemnation of larger exclusionary areas to ensure that all vented effluents would undergo more complete dilution before they reached the fenceline, improved physical integrity in radwaste holding systems, and higher stacks used to vent gaseous wastes. By such means, a proposed reprocessing plant at Hanford, Washington, with a radial exclusionary zone boundary almost ten times as great as Barnwell's, should reduce all exposures to about 40 percent of the dosages computed for organisms near the South Carolina site.

Power Parks and Radwaste Disposal

Although the reprocessing module of a power park could be designed to keep radioactive effluents at a tolerable level, even for organisms living in the immediate vicinity, it is scarcely believable that all effluents could be eliminated. Nor does minimization of emissions necessarily help solve the problem of long-term radwaste management—a particularly critical issue, since a reprocessing plant becomes a kind of depot for accumulating residues from reactors.

Short of successful development and introduction of fusion torches, the power park concept can offer no foolproof solution to the problem of radwaste disposal. However, in regions where geological conditions seems generally favorable for service as long-term depositories, the addition of a disposal-site capability could serve as a further requirement for approval of a proposed nuclear center. In this case, some power parks could provide not only for on-site recycling of usable plutonium and uranium but also for the ultimate disposal of any unusable wastes that might be generated within the park. The locale gaining the benefits of new employment and tax revenues as a result of power park construction would also get to keep the wastes there generated.

By no means would it be possible to co-locate fuel cycle facilities with every park. By no means would it be possible to have every co-located fuel cycle facility sitting atop a geological formation suitable for long-term mantle disposal or deep-mine storage of unrecycleable wastes. But where power parks could thus be sited at potential disposal areas, an additional measure of security might be brought to the nuclear system.

Regardless of whether or not radwastes might permanently be stored in the immediate locale of a given nuclear center, it would seem that concentrated siting of reactors would ease the problem of short- and intermediate-term storage prior to the shipment of radwastes for ultimate disposal. Particularly, the co-location of fuel fabrication facilities could improve the management of so-called low-level wastes. As noted in Chapter 3, half of all actinides by total weight remain, albeit in highly diffuse form, in the low-level wastes—especially in the residues of fuel fabrication. Steps to concentrate and discipline these wastes by confining them to a smaller number of sites could return significant dividends by keeping them in known locations likely to be frequented only by individuals with some awareness of nuclear operations and their potential hazards.

Because such wastes are diffuse and bulky, their segregation at a few sites would in time present disposal problems. Landfill areas nearby, capable of receiving enormous quantities of low-density radioactivity, might eventually be needed. This requirement adds to others that suggest the wisdom of searching for America's first power park sites in unpopulated areas. The potential impact of power park development on the social and economic conditions of rural regions will be considered in Chapters 6, 7, and 8.

Environmental Costs—The Power Park Heat Load

Adding to the economic costs of long-haul transmission are the main environmental costs of nuclear centers. Environmentalists have registered a variety of concerns: Will clusters of cooling towers offend the eye? Will clusters of reactors poison the site of a power park, perpetually withdrawing land from all alternative uses? Most students of the problem—opponents and proponents of nuclear centers alike—have concluded that one environmental issue emerges as truly critical. The heat dissipation requirements of a 20-unit power park raise two kinds of concerns: concern over the potential for *thermal pollution* of cooling water, and concern over the *atmospheric effects* of heat discharges to the airshed of a nuclear center.

Cooling Water Requirements

Since every megawatt of electricity implies a need to dissipate 2 megawatts in waste heat, the siting of a power station suggests a corollary requirement for a "heat sink" to which the thermal effluent may be rejected.

An impending water shortage, especially in western United States, operates as an increasingly powerful discriminator against light-water reactors in favor of energy technologies with higher thermal efficiencies (see Figure 5-5). Moreover, the Federal Water Pollution Control Act—the "Clean Water Act" of 1972—virtually prohibited the rejection of effluents, including heat, to the nation's waters after 1985.[19] In response to such restrictions, most new electric plants use wet cooling towers instead of flowing water.[20] Given the limits of temperature rise possible with the most efficient heat exchangers, a 1000-megawatt nuclear plant may need a million and a half gallons per minute circulating through its condensers. Most of this water will be vaporized in the wet cooling tower. Hence use of these devices implies an enormous consumptive use of water—with a corollary requirement for makeup water to keep them in operation. This requirement counts heavily against the feasibility of power parks. A 24,000-megawatt park would consume some 263,000 gallons of water per minute!

In the East, NRC researchers have found that most evaporated water would return to the watershed from which it had been withdrawn. So instead of speaking of "consumptive use" in such areas, it would be proper to think of power parks as accelerating the local hydrologic cycle by adding a further water-steam-rainwater sequence to the one that regularly occurs in nature. But in the arid West, makeup water might have to be imported by aqueduct to replenish the wet tower evaporation. Much of this imported water, after a single pass through the condensers of an electric plant, would leave the region as wind-borne evaporate and therefore become unavailable for recycling within the watershed.

A more desirable option from the standpoint of water conservation would involve the use of dry towers, in which water is evaporated and then recycled within the cooling system of the plant. However, NRC analysts estimated that nuclear centers in remote regions of the West using dry towers, though technically feasible, were probably ruled out on economic grounds.[21]

Special ponds or channels, sometimes supplemented by spray devices to increase the water's surface area, can perform the same function as wet towers do—dissipation of heat by evaporation. In place of capital expenditures to build

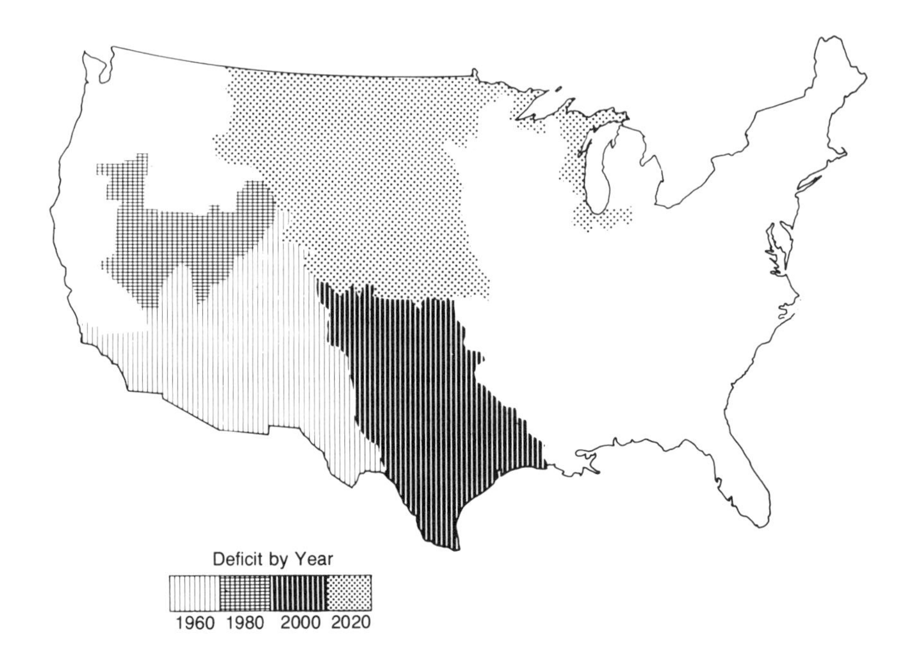

Source: Adapted from Ronald Ridker, *Population and the American Future* (U.S.G.P.O., 1972), p. 46. Estimates assume rapid economic growth, maximum development of water storage facilities, and tertiary treatment. Reprinted by permission.

Figure 5-5. Water Deficit Regions in the United States, 1980-2000.

Notes to Figure 5-5. Present and projected limitations of available cooling water could prevent the development of nuclear centers in some areas west of the Missouri River *unless* (1) sites are identified quickly, and nearby waters reserved against alternative uses, or (2) major projects for interbasin transfers of water are initiated.

For example, Great Lakes water might be carried by aqueducts to nuclear centers in the Dakotas or Nebraska. Salt water from the Gulf of Mexico could be brought to power parks in Colorado. The latter possibility would invite efforts at the development of desalination plants as integral elements of agroindustrial complexes centered on the parks. In the most grandiose scheme yet proposed, huge aqueducts would bear the waters of Canadian rivers from British Columbia thousands of miles through the Rocky Mountains—first to industrial plants and thence to the irrigation fields of southwest United States and Mexico.

In some circumstances, long-haul transport of cooling water could be achieved more cheaply than transmission of electricity over equivalent distances. Transport of water at an estimated $5 per kilowatt of capacity per hundred miles carried may prove less restrictive an economic constraint than the layman might have supposed. For certain projects, the corresponding figure for long-haul transmission of electricity is $14. (See pp. 33-40, G.E. Preliminary Draft of the N.S.F.-funded study, "Evaluation of Energy Parks versus Dispersed Sites," November 15, 1974, citing Chilton's "Cost Engineering in the Process Industries.")

the massive reinforced concrete towers, ponds and channels require major land purchases—typically one-half acre underwater for every thermal megawatt of electric capacity being serviced.

The in-residence use of water in the mid-1970s averaged about 75 gallons per capita per day across the United States. By contrast, the water needing evaporation in order to produce electricity is on the order of 10 to 15 gallons per capita—a figure that will increase as electricity is substituted for less water-intensive energy forms, and as the American economy itself becomes increasingly energy intensive. Yet even with significant increases between now and the year 2000, these water-use figures suggest that so long as vapor from cooling towers does remain within the watershed, withdrawals may be regarded not as a net loss but as an epicycle within the region's hydrologic cycle whereby a fraction of the residential commitment is reused to bear off waste heat. This is not, of course, to deny that unforeseen side effects may result in particular regions from such disruptions of the natural cycle. In this respect as in most others, the power park strategy presumes gradual development and careful planning in the light of accumulating data on the balance of advantages and disadvantages.

The Atmospheric Heat Load

With or without water losses to the regional watershed, evaporative cooling—or, for that matter, any other heat rejection technique—results in major transfers of warm air to the local airshed. Indeed, most studies of nuclear centers suggest that, by all odds, the gravest environmental problem posed by a power park could lie in a climatologically disruptive "heat island" effect.[22]

A nuclear complex the size of the proposed River Bend, Louisiana center (about 33,000 megawatts by 1992, drawing cooling water from the nearby Mississippi) would vent more waste heat to the local atmosphere than the city of Chicago releases to the air south of Lake Michigan. For a comparison with natural phenomena, this heat discharge about equals the energy flux needed to form a large snow cloud. In certain climatological conditions, heat releases on this scale could conceivably make the difference between marginal disruptions (i.e., increased local precipitation downwind of the park's cooling towers) and the occasional triggering of major weather events. A nuclear center on the prairie, capped from time to time with minicyclones, suggests the kind of concerns that could decisively tip opinion against power parks in climatologically volatile regions.

Weather experts have established that severe climatological modifications are highly unlikely. (Actually, the heat discharge of a power park, though large by human standards, is orders of magnitude smaller than the energy content of a tornado.) But the possibility of more limited atmospheric effects—the euphemism is "storm enhancement"—remains. Increased local precipitation must be expected near power parks cooled with evaporative towers.

Some regions with especially unstable climatic conditions would be "red-lined" by energy park planners, as would areas known to be earthquake-prone (see Figure 5-6) or severely short of water. These regions would then become major exclusionary districts.

Perhaps the ultimate thermal effects of nuclear centers can be established neither by theoretical analysis nor by scale-modeling of heat-release phenomena. Perhaps definitive answers on the climatological effects of power parks will emerge only as information accumulates over the years. Until better data are available, however, careful spacing of units within the fenceline seems the key to the safe dissipation of waste heat. The best scientific opinion available to NRC analysts suggests that most adverse climatological effects may be prevented by deploying quads of reactors at least two-and-one-half miles from one another so as to provide ample spacing for the dissipation of released heat. (Virtually all the economies of concentrated siting, such as opportunities for modular construction and special assembly facilities, hold at the level of a two-and-one-half-mile spacing pattern. Moreover, NRC researchers found that situating quads of reactors this distance from one another would reduce the maximum radiological exposure of an organism living fulltime at the power park fenceline to a level below that associated with the same number of nuclear units dispersed across a large region.[23])

A cautious response to climatological and seismic considerations suggests the desirability of blocking out areas of severe weather concern on a map of the United States and then further excluding regions of known seismic instability. The NRC study team used such a procedure for seismic exclusion and then added overlays to eliminate areas of water shortage and high population density. Also excluded from consideration as sites were certain public lands and national parks areas.

Summary: The Case for Power Parks

Any significant addition to the nation's electric generating capacity between now and the year 2000—whether new units are sited in a series of power parks or in a multitude of dispersed locations—will severely intensify competition for scarce capital funds. Hence the economic case for parks hinges on the ability of nuclear centers to return substantial savings compared with the cost of installing equivalent megawattage in the dispersed mode. Power parks, then, offer a partial solution to problem number 1 in the nuclear policy field—the problem of escalating capital costs.

Power parks also offer a partial solution to problem number 2—the problem of radioactive hazards. By closing the fuel cycle, by providing an economic base

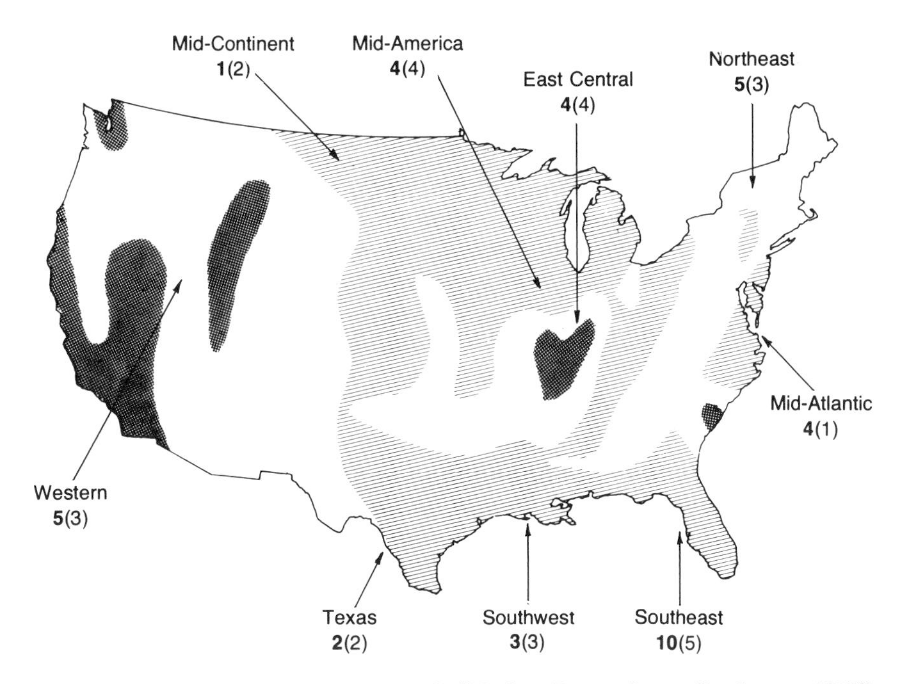

Source: Adapted from Appendix A, Part 1, "Nuclear Energy Center Site Survey–1975" (U.S. NRC, January 1976).

Figure 5-6. Projected Electric Demand versus Seismic Suitability.

Notes to Figure 5-6. Boldface figures indicate the maximum number of nuclear centers that could, at the upper limit of a national commitment to the concept of power parks, be supported by 1975 projections of future electric demand in each of the nine U.S. Electric Power Reliability Council regions. The parenthesized figures give the corresponding number of power parks per Reliability Council region, assuming a target total of approximately 27 nuclear parks, to meet two-thirds of the 800,000 megawatt nuclear capacity needed in the year 2000 assumed in Figure 2-1, leaving the remaining one-third of nuclear demand to be served by reactors on dispersed sites.

The blacked-out areas mark regions of sufficient seismic unreliability to be redlined as nuclear center sites. This does not absolutely preclude power park siting in such areas. Rather, it suggests that the added costs to "harden" facilities in order to provide an extra margin of safety in the event of major local faulting would be excessive.

Preliminary geological surveys conducted as part of the NRC national site survey suggest that the shaded areas are, at least from the seismic standpoint, likely regions for nuclear center development. More detailed geological studies of specific proposed locations would be needed to determine the numbers of suitable sites in the remaining areas.

From the standpoints of seismic suitability, water availability, land costs, and population distribution, the NRC analysts concluded that adequate numbers of power park sites can be found in every one of the main U.S. Electric Power Reliability Council regions.

In terms of seismic stability, the central landmass seems the most likely region for nuclear center development. But the geographic distribution of the projected national electric load points toward growing demand along coastal strips. Obviously, it would be desirable to emphasize research toward any breakthroughs in low-cost transmission that would permit nuclear centers to be sited on the central plains, generating power for export to the continental rim.

to support special cadres of nuclear qualified workers, and by reducing the number of waste-production points—in these ways, nuclear centers can help reduce the risks of atomic fission.

Table 5-2 shows the savings that might be realized by a transition away from the siting of reactors in quads near electric load centers to siting in power parks. Savings of up to 10 percent may be achieved in the production costs of fission-based electric power by clustering reactors, assuming a 13 percent saving in capital costs and a *tripling* of transmission costs. The center column of figures suggests that with only a 9 percent reduction in capital outlays and as much as a fivefold increase in transmission costs, power parks can still break even with the costs of electricity from nuclear plants sited in the dispersed mode. But even at this break-even level, parks would still imply certain nonpecuniary benefits, such as reduced risks of plutonium diversion, more favorable occupational accident levels, and improved radwaste management.

Table 5-2
Additional Social and Economic Savings from Nuclear Centers

	Mills per kwh		
		Nuclear Centers	
	Dispersed Quads	*Case No. 1*	*Case No. 2*
1. Capital Cost of Plant	25.1	22.95[a]	21.85[b]
2. Transmission, Distribution	1.1	4.5	3.3
3. Fuel Cycle Costs			
Fuel Preparation	2.35	2.35	2.35
Fuel Recycle			
Spent Fuel Transport	.15	.02[c]	.02[c]
Reprocessing (w/Fuel Credit)	.1	.1	.1
Inventory Charge	1.25	.65	.65
4. Environmental Costs	.503	.503	.503
5. Other External Costs			
Sabotage and Diversion	.053	.026	.026
Accidents, Explosions	.002	.001	.001
Occupational Hazards	.01	.009[d]	.009[d]
Low-Level Radiation	.001	.001	.001
6. Transport, Disposal of Wastes			
Radwastes (High Level)	1.06	.53	.53
Low-Level Wastes	.1	.1	.1
	31.779	31.740	29.440
Gain from Nuclear Center Configuration			3.339

[a]At 9 percent capital cost savings.
[b]At 13 percent capital cost saving.
[c]Ref. p. 5-110, USNRC Nuclear Energy Center Site Survey.
[d]Ref. p. 5-30 *ibid.*

What do the cost estimates that have emerged thus far suggest regarding the economic gains that might be achievable over the next generation or so from a full Promethean progression? As shown in column 2 of Table 5-3, the phasing out of dispersed fossil-fired units in favor of dispersed nuclear plants could cut electric costs by almost 2.5 mills per kilowatt-hour (again, taking mid-1970s dollars as a common base). Column 3 suggests that if this transition were complemented by a shift from dispersed nuclear siting to the location of two-thirds of all future fission-based generating units in parks, an additional saving of almost 4 mills per kilowatt-hour would be possible. Finally, the estimates in column 4 show that the incorporation of all new fusion-based electric power after the year 2000 into power parks might reduce customers' electric bills by yet another 12 mills per kilowatt-hour.

Of course, both technical and institutional obstacles combine to prevent any direct transition from a predominantly fossil-fired dispersed electric generating industry to a system based on fission and fusion plants in power parks. The Promethean progression must be negotiated, if at all, in more or less discrete phases—first, through the development of fission-based power parks between now and the year 2000, followed by the introduction of CTRs as rapidly as possible in the twenty-first century. Chapters 6, 7, and 8 outline a strategy for managing such a process of long-term change.

Table 5-3
Gains Theoretically Attainable with a Full Promethean Progression

	1	*2*	*3*	*4*
	Base Case: Dispersed Coal-Fired Plants[a]	*Positive or Negative Change in Cost of Electricity (Mills per kwh) with Shift*		
			Nuclear Centers	
		Dispersed Fission	*Fission*	*Fusion*
Capital Cost of Plant	18.26	−6.74	+3.15	+2.2
Transportation, Distribution	1.1	–	−2.2	+2.2
Fuel Cycle Costs	9.65	+5.6	+ .93	+3.1
Environmental Effects	5.94	+5.437	–	+ .103
Other External Costs	.2	+ .134	+ .029	+ .034
Transport, Disposal	–	−1.16	− .53	+ .53
			+1.379	+8.167
Cumulative Saving		2.471[b]	3.850[c]	12.017[d]

[a]Figures taken from column 1, Table 2-1.
[b]Based on estimates in column 2, Table 2-1.
[c]Based on case 2, Table 5-3.
[d]Extrapolated from comparison of columns 1 and 2, Table 4-2.

Notes

1. For an overview of the concept, see Seymour H. Smiley et al., "Feeding the Glutton: The Trick is to get the nuclear energy center 'out of town' . . . ," *I.E.E.E. Spectrum* (July 1976), pp. 74-83.

2. The studies of agroindustrial complexes, all completed at Oak Ridge, included the following, released in November 1968: "Nuclear Energy Centers: Industrial and Agro-Industrial Complexes," ORNL-4290; "Potential Agricultural Production from Nuclear-Powered Agro-Industrial Complexes Designed for the Upper Indo-Gangetic Plain," ORNL-4292; and "Steel-Making in an Industrial Complex: Acetylene Production, etc.," ORNL-4294.

3. Comments on the achievements of Alvin Weinberg and the controversy that continues to surround his work appears in Daniel Jack Chasan, "A Pioneer Feels the Atom is Still Our Best Answer," *Smithsonian*, 82-89. The following papers give a representative sampling of Weinberg's positions: "Limits to the Use of Energy," with R. Phillip Hammond, 58 *American Scientist* (July August 1970), 412; "On the Siting of Nuclear Reactors" (AEC discussion paper, December 17, 1971).

4. Part II., USNRC *Nuclear Energy Center Site Survey–1975* (USNRC, Bethesda, Maryland, NUREG-0001, January 1976, p. 8-2.

5. Center for Energy Systems of the General Electric Corporation, "Assessment of Energy Parks versus Dispersed Electric Power Generating Facilities" (May 30, 1975). The G.E. study, which runs to thousands of pages, is replete with figures and systematic quantitative evaluations. I have drawn extensively from this reference document, which is henceforth cited as "G.E. Assessment."

6. Pub. L. 93-438, 88 Stat. 1233.

7. U.S. Nuclear Regulatory Commission: NUREG-0001. Henceforth cited as "NECSS."

8. Ibid., Part I, pp. 3-1ff.

9. See generally the Berlin et al. study for the Ford Foundation Energy Policy Project, *Perspective on Power*, especially Chaps. 5-6.

10. G.E. Assessment, Vol. I, p. ES-7, and Vol. II, Chap. 9.

11. Ibid., Vol. I, p. 4-296.

12. NECSS, Part I, p. 3-15.

13. G.E. Assessment, Vol. I, pp. 4-285ff.

14. The figures used in this analysis are somewhat more conservative–that is, potentially disadvantageous to power parks–than those used by the NRC. See NECSS, Part III, pp. 4-32ff, and especially at p. 4-39.

15. Ibid., pp. 4-64ff.

16. G.E. Assessment, Vol. I, p. 4-140.

17. See generally "Report of the Ad Hoc Study Group on Integrated versus Dispersed Fuel Cycle Facilities," F.K. Pittman, Director (U.S. Atomic Energy Commission, October 1974), p. 64.

18. Ibid., p. 66.

19. 33 U.S.C. 1251 *et seq.*

20. For summary figures on aspects of the cooling problem, see G.E. Assessment, Vol. II, pp. 7-12ff.

21. See NECSS, Part I, p. 4-12.

22. A comprehensive quantitative study of perhaps the most serious potential environmental problem of energy parks—the "heat island" effect—will be found in L. Randall Koenig and C.M. Bhumralkar, "On Possible Undesirable Atmospheric Effects of Heat Rejection from Large Electric Power Centers" (The Rand Corporation, December 1974), R-1628-RC.

23. NECSS, Part I, pp. 10-8, 10-15. See also at p. 3-15.

6 The Challenge of the Intermetropolitan Frontier

The concept of nuclear centers withstood increasingly intensive study through the early 1970s. A clear momentum seemed to be developing, as the Federal Energy Agency contracted for studies of potential "designated sites" for major power-related developments. Large tracts near Harbor Beach, Michigan, at Glasgow Air Force Base, Montana, in the Tularosa Basin, New Mexico, and at Camp Gruber in Oklahoma have been surveyed as possible sites of power parks.[1]

The most conspicuous example of ongoing power park evolution centers on the coal-fired electric plants at Four Corners (where Arizona, Colorado, New Mexico, and Utah meet). These plants, with some 4000 megawatts of rated capacity, generate electricity for export to southern California. Already, the Four Corners installation is the one human artifact visible to orbiting astronauts. Yet the total capacity envisioned for the region may ultimately top five times the present figure. In general, residents of the area favor further development, provided that measures are taken to ensure retention of economic benefits within the region.

Full development of the Four Corners region will make an entire multi-county area into a kind of super power park, fueled by the rich coal deposits of the Southwest. Unfortunately, there is little evidence that the electric capacity will be brought on line as part of a long-range plan to ensure the regional economic development on which local support for the utilities' designs are based.[2]

Rich strip-minable coal seams also underlie parts of Wyoming, Montana, and the Dakotas. In 1971 the North Central Power Study, a planning effort funded by a consortium of 27 utilities and the Bureau of Reclamation, recommended installation of some 43,000 megawatts in mine-mouth generating capacity near Gillette, Wyoming.[3] This once sedate cattle town would become the center of a major power park extended over 200 square miles. By this plan, coal could be stripped, hauled short distances to electric stations, and then effectively exported via high-voltage lines to Minneapolis, Omaha, St. Louis, and other load centers to the east and south.

Seven hundred and fifty miles or so northwest of Gillette, plans have been developed to augment the already considerable nuclear capabilities located in Washington state with a new 30,000-megawatt complex at Hanford.[4] And 750 miles to the northeast, a team of researchers from the University of Minnesota Center for Urban Studies recommended a feasibility study for a power park to be sited in Aitken County, an hour's drive north of Minneapolis.[5]

Plans have also proceeded apace in the East. In 1974 three Pennsylvania utilities–Pennsylvania Power and Light, the Philadelphia Electric Company, and the General Public Utilities Service Corporation–combined to evaluate sites for a jointly owned park. From 10 to 20 megawatts would be generated, half by coal and half by nuclear plants, for the Philadelphia market. The following passage from a 1974 report by the Pennsylvania study group suggests utility executives' reading of trends in the industry:

> The larger energy parks being considered in this report appear to be a natural progression from the "first generation" parks currently in operation, construction or planning. Keystone, Conomough and Hoover City are really energy parks. Another example is the Peach Bottom, Muddy Run, Conowingo and proposed Fulton complex which will represent nearly 6000 MW of capacity along a 13 mile stretch of the Susquehanna River. Four Corners, Oconee, Shearon Harris and Greenwood are more sophisticated parks.[6]

The record suggests that public reluctance to accept power parks rather than economic impracticality or technical infeasibility sets the limit on plans to move forward. While the Pennsylvania project was shelved for further study following an adverse reaction by citizens in builtup areas thought to be prime sites, another park in neighboring New Jersey is developing more or less in the way that Topsy "just growed." When plans for a nuclear station on the Delaware River near Trenton had to be reconsidered under pressure from environmentalist opponents, the New Jersey Central Power and Light Company shifted the proposed location of the reactors to an already approved nuclear site lower on the river. Eventual addition of the two displaced plants to the two already under construction near Salem, New Jersey could bring total capacity in the vicinity to upwards of 5000 megawatts. If this process of siting-by-convenience were to continue, Salem would in time itself become a nuclear complex–but by happenstance and gradual accretion of new capacity rather than through foresighted planning.

(Power parks have also been proposed that would use solar energy as well as nuclear or fossil fuels. In the most thoroughly thought-out plan, advanced by Aden and Marjorie Meinel of the University of Arizona, some 5000 square miles in the desert areas of Arizona and California would be dedicated for use as "solar farms"–covered with mirrors to collect radiant energy for electricity to power a new linear city running from the Mexican border along the Colorado River almost to Lake Meade in Nevada.[7])

Evidence from all points of the compass leaves little doubt, then, that America *will* have power parks. Will utilities be permitted to build ever bigger stations simply because such a course best satisfies the logic of the corporate dividend? Will they be permitted to add new reactors on existing sites merely to avoid full regulatory review of proposals for new dispersed sites? Or will power parks be carefully planned to ensure both representation of all potentially interested groups and the realization of national socioeconomic goals?

Power Planning and the Sociological Imagination

An effective national policy to govern the development of power parks will require a shift in planning philosophy. Traditionally, the intimate links between social values and the growth of electric loads attracted scant interest from utility executives, who had been concerned merely to ensure that calls for more power could be answered—and regardless for what purpose, from driving electric toothbrushes to lighting the streets in America's inner cities. The mechanisms whereby social phenomena translate into rising or falling demand for electric power evoked only slight curiosity in the energy community. American society was seen as "externalized" from the utility industry that served it. The dynamics of social change were treated as if they existed in some kind of black box, detached from the utility planner's or the government official's purview.

In this traditional, detached view, the driving forces of economic growth and population distribution attracted the planner's concern only insofar as they translated into changes in the demand for power. What occurred within the black box of society interested the utility executive only if it pointed directly to an upward or a downward inflection in the projected rate of electric growth.

The aim of providing adequate low-cost electricity without distraction by "extraneous" factors—such as the socioeconomic or environmental impacts of plant expansion—has narrowed the perspective of utility decisionmakers. The most serious consequence of this, as of any narrowing of perspective, lies in the danger that major opportunities will be missed because the decisionmakers' limited viewpoints blinds them to their existence.

In *The Sociological Imagination*, C. Wright Mills wrote of that outlook by which the individual begins to relate his or her own situation to the seamless historical fabric of which personal circumstances form an integral part.[8] A sociological imagination further implies a sensitivity to the linkages by means of which a decision in one subject area will ramify to others. Power park development demands analysts who possess sociological imaginations. In place of fragmented policies formulated by narrow specialists in such subjects as energy supply, economic growth, manpower development, and so forth, a single policy should be formulated with an eye to the interacting parts of the social system.

A power park cannot exist decoupled from its social context. The commercialization of a new technology—and nuclear power remains that—creates externalities or spillover effects. These effects influence the welfare of individuals other than the buyers and sellers. General Electric or Westinghouse may sell, and a private utility may buy, a nuclear reactor, but not without indirectly influencing thousands of individuals concerned about radioactive releases, land usage, and aesthetic degradation of the locale.

All those dimensions of nuclear operations which imply consequences in addition to the provision of electric power may be thought of as interfaces between the industry and society. To evaluate the contribution of a power park to electric supply, while failing to evaluate the socioeconomic impacts of the new generating station, risks doing positive harm to the quality of social life. Planners must identify important interfaces, then internalize them as primary factors in the siting and design of nuclear centers. Instead of focusing exclusively on the question, How much energy need be produced to meet market demand?, planners should emphasize the way in which energy production and distribution are accomplished. Instead of merely trying to meet demands registered by consumers within a black box, they should seek to make positive use of the multifold interfaces that couple the energy industry to society.

A Demographic Approach to Power Park Planning

In 1972 members of the National Commission on Population Growth and the American Future advanced four social goals:

1. To promote a high quality urban environment.
2. To promote a range of lifestyle options.
3. To ease problems of population movement.
4. To foster free choice of residential location.

These goals point to a set of social problems in late twentieth century America that may be subsumed under a "demographic approach" to power park planning.[9]

The second and fourth goals of the Population Commission suggest a need to create different kinds of urban centers. The creation of new towns and the stimulation of old ones could contribute, at least marginally, to a reduction in the growth rate of the biggest urban areas in which, by the year 2000, five out of every six Americans will probably reside (see Figure 6-1). In area, these 24 centers should account for some one-sixth of the continental United States.[10]

(It should be emphasized that the size distribution of cities in a modern society tends to be notably stable over long periods. Massive and perhaps coercive interventions would be needed to effect major changes in the demographic patterns of American settlements—and the resulting social dislocations could make such efforts self-defeating. Hence demographers see no way to reverse metropolitan growth. At best, a policy of "growth stimulation" aimed at attracting people from big cities to smaller population centers in the intermetropolitan hinterland could, while helping to revivify the frontier, only *slow* the relative growth of major metropolitan centers.)

The history of American urban growth[11]—matched, in the twentieth century, by rural decline—gives perspective to the case for using power parks as sources of economic stimulation on the intermetropolitan hinterland.

Source: Adapted from *Population Growth and the American Future* (U.S.G.P.O., March 1972), Fig. 3.3, p. 33. Reprinted by permission.

Figure 6-1. Projected American Urban Centers, The Year 2000.

The Coming of the Dual Society

In the late eighteenth century, leading American towns began to serve as investment outlets and banking centers for funds that had accumulated in commercial agriculture. America's first growth cities served both as sources of capital to be invested in the development of the frontier and as staging bases from which pioneers could jump off for the continental interior. The mineral resources of that interior underwrote the next phase of American development. The coming of this phase was reflected in a shift from the undisputed priority of agriculture to a new kind of prosperity based on agriculture *and* industry. From the 1830s onward, the industrial Northeast developed to encompass the

Pennsylvania coalfields and Lake Superior's iron ores as well as the financial, entrepreneurial, and specialized manufacturing centers of the older Eastern cities.

In time the plains–those in grain as well as those still farther West in forage for cattle–began to produce staples in demand on national and world markets. The growth of America's depot cities depended on the richness of the hinterlands accessible to their merchants. Deepwater ports developed to serve manufacturing centers and agricultural hinterlands ever farther removed from the Eastern littoral. The typical urban growth center became a port on a major oceanic or riverine trade route. Baltimore, New York, and Philadelphia were cities of the first type; bustling new centers, such as Cincinnati, St. Louis, and Minneapolis, were cities of the second.

The elaboration of a national transportation network based on rails and roads helped tie supply areas to manufacturing centers and nodes of commerce. In time too, easy air and roadway transport facilitated the growth of "footloose industries," such as research firms and companies dealing in high technologies. Such firms need not stay tied to specific regional bases by natural resource deposits.

The rise of footloose industries, together with the growing demand for leisure and retirement opportunities in the Southern and Western sun belts, helped break the old pattern of Northeastern dominance. Tens of millions of Americans since World War II have swelled the rolls of relatively new metropolitan areas–areas outside the old agroindustrial powerhouse anchored to New York, Pittsburgh, and Chicago. The urban growth crescent of late twentieth century America runs from the burgeoning cities of the Southeast (Miami, Tampa, Atlanta), through the modern Texas and Arizona boom towns (Dallas, Houston, Phoenix), up through California's megalopolis (San Diego to San Francisco), to Portland and Seattle. Centers of economic and cultural activity now exist in all quarters of the land. Birmingham is every bit as much a national industrial city as is Pittsburgh. The San Francisco metropolitan area rivals New York as a cultural and intellectual capital.

By the same sign, energy loads of significant, and prospectively of fast-growing, dimensions are to be found through the country. Whether electric demand is to be met by power parks or by dispersed generating units, the national power grid must increasingly reflect the national dispersion of population. This dispersive trend pulls the pattern of future capacity out farther from the heavily industrialized Northeast to what were once Southern and Western frontier areas.

In 1975, census data showed a population increase of more than 8 percent in the South and of almost 9 percent for the 11 Western-most states, as against about 2 percent for the Midwest and only 0.8 percent in New England and the mid-Atlantic states. Overall, in the first half of the seventies the South and the West combined grew at about twice the national rate.[12]

But if the number of major cities has increased since World War II, population densities in most of those urban areas have actually decreased since the mid-fifties as a result of middle-class flight—first to relatively close-in suburban tracts and then to overspill acreage 25, 50, even 75 miles from downtown centers. These sprawling fringes define the modern "conurbation," a term coined by European urbanologists for the gradual interpenetrative buildup of populations in the intervening spaces between two central cities.

Studies of Minneapolis' metropolitan growth by the Minnesota State Planning Agency suggest the outreach of the center city in metro-America. Before the development of the Twin Cities' modern highway system in the 1960s, a typical resident of Minneapolis or St. Paul could travel from point to point within an area of some 800 square miles in an hour's time. By 1985, almost 6000 square miles will be within one hour's commuting time—an eightfold increase, while population multiplied only a little more than three times.[13] As will be discussed in Chapter 8, similar regional roadway developments, centered on power parks in rural areas, will be necessary to develop cities of the second sun.

Beyond the commuting outreach of the main cities lie America's villages, hamlets, and rural areas. As shown in Figure 6-2, surprisingly large areas of the United States suffered population declines between 1960 and 1970. The intermetropolitan frontier is everything that the conurbation is not—economically stagnant rather than dynamic, declining in population rather than growing.[14]

From the 1930s onward, mass outmigrations of the agrarian poor have made for shrinking rural markets and shrinking workforces. Table 6-1 shows the demographic structure in a typical regional hierarchy of American settlements. In general, the rates of population growth and economic activity are strongly positive for the already-big communities. But in the smallest settlements, these rates have inflected downward. Or at least population growth in villages and hamlets has lagged behind that in settlements of larger size. Across the prairies of the upper Midwest, the prospects of America's villages diminish as young people, finding inadequate opportunity at home, move to the cities.

America has split into a dual society—a mushrooming urban population and a residual population in an intermetropolitan hinterland marked by economic stagnation, diminishing opportunity, and lagging population growth rates. The trends toward metropolitan gigantism on one side and intermetropolitan stagnation on the other suggest the need for a new planning framework. Such a framework should link approaches to America's future population distribution with a coherent economic growth policy, and relate both of these to the national energy supply system. How might a national policy for electric power plant sighting contribute to the goals announced by the Population Commission?

The answer is, by contributing to the fulfillment of trends and aspirations that are already discernible in the reaction of the American people to the

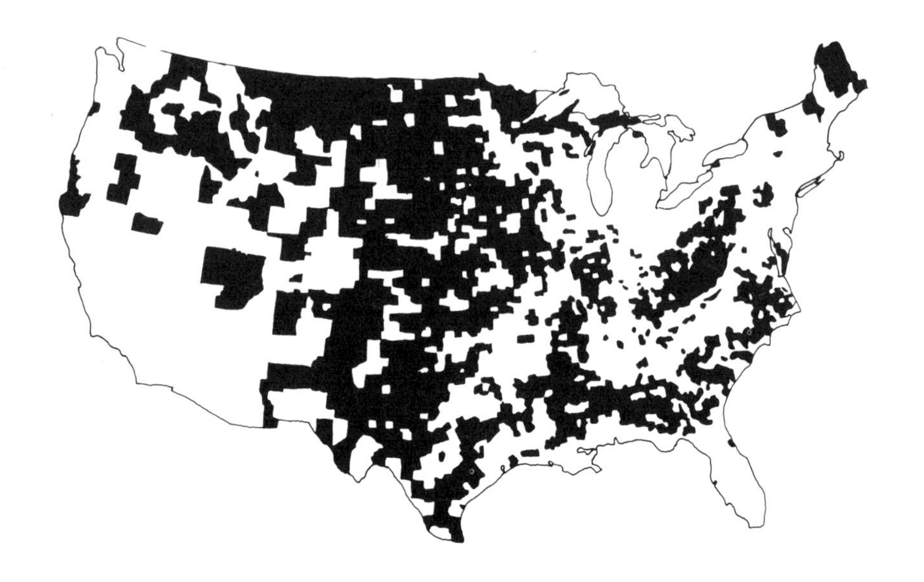

Source: Adapted from *Population Growth and the American Future* (U.S.G.P.O., March 1972), Fig. 3.2, p. 28. Reprinted by permission.

Figure 6-2. Areas of Population Decline between 1960-1970.

Notes to Figure 6-2. Significant bands of the country had populated declines in the years 1960-1970. Many of these areas, indicated in black, lie beyond the metropolitan outreach. Such stagnating frontier regions offer likely sites for the development of power parks to stimulate economic development and population growth.

Central to the case for using power parks as economic growth poles is the fact that new family formation, a corollary of the "bulge" of population in the child-bearing age groupings, implies disproportionately large requirements for new houses, new cars, new schools–and hence for new energy development.

The National Commission on Population and the American Future projected that an additional 71 million Americans would be alive in the year 2000, over and above the population of 200 million in 1968. Should the average natality rate correspond to a three-child family, the incremental addition would come to 121 million by the turn of the century.

The U.S. Census Bureau in 1975 revised these figures downward somewhat–to a probable total population of at least 250 million in the year 2000. (See Table O, Series II-L,R,X, *Population Estimates and Projections*, October 1975, U.S. Department of Commerce.) This lower estimate results mainly from a reduction in fertility from previously expected levels, compounded by an apparent trend toward delaying the age of women at which children are, on the average, born.

Over the entire range of different assumptions used in the Census Bureau's 1975 projections, the rate of population change goes negative only in one boundary case, and even then, not until the second decade of the twenty-first century. (See Figure 2, Series III, *Population Estimates and Projections.*)

In such a context, new jobs must be found for men and women entering the labor force in order to provide for them *and* for their nonproductive dependents, both young and old. A significant percentage of the needed new employment opportunities may be generated in well-planned cities of the second sun.

Table 6-1
The Urban Hierarchy in Upper Great Lakes States

	Demographic Characteristics	
	Percent in Decline	*Median Population*
Viable City-Types		
Metropolis	0	1,4000,000
Wholesale-Retail Center		
Primary	14	55,000
Secondary	10	32,000
Complete Shopping	12	7,000
Partial Shopping	20	2,500
"Convenience Trade"	21	1,500
Not Viable		
Partial convenience-Village	42	800
Hamlet	77	200

Source: John R. Borchert, *Upper Midwest Changes in the 1960's*, Upper Midwest Research and Development Council, 1965. Cited with further supporting data in the Upper Great Lakes Regional Commission Study, *Growth Centers and their Potentials in the Upper Great Lakes Region*, May 1969. p. 32. Reprinted with permission.

problems of the metropolis. Power parks may be used not merely to provide cheaper, safer electric power but also to help give effect to Americans' growing preference for a changed demographic complexion in their future society. Such a conception would underlie the use of nuclear centers as economic growth poles on the nonurban frontier. Such a conception, too, gives rise to the vision of American cities of the second sun.

Reversing the Decline of the Frontier

The most useful concept uncovered by the National Commission on Population and the American Future was that of *size of place*. If an analyst knows the population of a given settlement, from a hamlet up to a megalopolis, he or she can infer more additional information about the quality and pace of life within that community than is possible from any other piece of information. Is a given settlement small or large? Towns and villages, which generally show relatively low levels of economic turnover, are frequently also marked by declining populations, and sometimes by social provincialism. But small towns also offer the satisfactions of an easier pace of life, pollution-free air, lower crime rates, and a sense of community not found in the metropolis.

Big cities imply high levels of economic activity. They imply rapid population growth. But of course too, big cities today suggest increasing levels of crime and congestion. The image of the town or small city, and the symbol of the urban dynamo, bracket the consciousnesses of increasing numbers of Americans. The interaction of town and city within a given region, we shall see, also may hold the key to rejuvenation of the intermetropolitan frontier.

Reacting against the decay and frustration of the sprawling conurbation, Americans began in the late 1960s to reveal a preference toward relocation to smaller cities. Opinion surveys sponsored by the Population Commission in the late 1960s, when the megalopolis appeared to be growing more quickly than ever, showed a striking nostalgia for smaller communities (see Table 6-2). By the mid-1970s, hard statistics were available to suggest that Americans' preference for the more peaceful, simpler lifestyles of smaller cities represented more than the unrealistic fantasies of harried urban dwellers. For perhaps the first time in American history, a positive rate of population growth in specific rural areas—parts of Appalachia and the Ozarks, in the Rocky Mountain region from Idaho to northern New Mexico, across the Upper Great Lakes—were matched by gradual declines in urban concentrations. Not just the crumbling central cities but several major metropolitan areas taken as wholes began to show declining rates of population growth.

Between 1960 and 1970, 6 million Americans migrated from nonmetropolitan areas to urban centers. By contrast, between March 1970 and March 1974, U.S. Census Bureau figures showed almost 6 million moving *out* of metropolitan areas, while the urban influx fell to slightly more than 4 million. Rural counties gained 3 percent nationally from 1970 to 1973, compared with population losses averaging between 4 and 5 percent during the sixties.[15]

The return of Americans to medium-sized settlements on the frontier would presumably have been even more pronounced if family heads might have selected their communities on the basis of true residential preferences rather than on the basis of economic necessity. But the continued dominance of metropolitan areas as job centers dampened the trend toward population

Table 6-2
Residential Preferences of Americans, by Size of Community

	Percent of Total Respondents	
	Where Do You Live Presently?	*Where Would You Prefer?*
Urban Residence		
Big City or Suburb	27	14
Medium City or Suburb	28	22
Small City or Town	33	30
Open Country	12	34
	100	100

Source: Report of the Commission on Population Growth and the American Future, *Population and the American Future* (U.S.G.P.O., March 27, 1972), p. 34.

redistribution. Even when industries fled the central cities, they often moved to sites within the metropolitan commuting perimeter. Because the mobility of their employees continued to be limited by job availability, homeowners who worked in such firms moved farther out into suburbia. But they necessarily remained, along with their employers, within daily commuting distances of the old urban cores.

A sustained trend toward population relocation into rural areas–however much such a movement may be desired by thousands of Americans–therefore seems constrained by the relatively bleak prospects of generating sufficient economic activity in the intermetropolitan hinterland. Needed is a policy that will permit increasing numbers to opt for productive lives outside the largest metropolitan centers.

Nuclear centers have a major role to play in a national strategy aimed at developing such economic opportunities. When introduced into stagnant regions on the intermetropolitan frontier, power parks could stimulate economic growth through three kinds of effects: (1) *concentration effects,* whereby revenue-producing activities that are normally dispersed throughout the nation might be focused within the region; (2) *export effects,* whereby continuing demand for electricity in the national market might lead to a spiral of new investment in a region that hosts a nuclear center; and (3) *agglomeration effects,* whereby economic activity in the region might be stimulated by local opportunities to achieve efficiencies which are made possible by careful location and design of the power park.

Power Parks and Regional Development

To establish an economic basis for development of the frontier implies an effort to alter the structure of the regional economy. That structure may be thought of as a hierarchy of settlements from the villages which to this day retain the spirit of rural America up to the major metropolitan concentrations.[16] Typically, the number of centers in each category decreases with an increase in the size of settlement type: many hamlets, a few medium-sized cities, and one metropolis enjoying regional primacy.

Hamlets and villages provide minimum commercial services for residents of the surrounding districts. Larger towns and cities produce some manufactured goods. They also provide the more specialized services demanded by citizens of the region but which cannot be supported within the markets served by villages. In the prime city, production of the widest range of goods and services helps support employment throughout the region. Salespersons, repair workers, delivery men, and other operatives must market and maintain the products of the prime city. Often too, firms locate in outlying towns to supply semifinished products needed by the manufacturers in the prime city.

Economic activity throughout the region, then, depends on reciprocal contributions by the leading city and its satellite settlements. It follows that the recovery of a stagnant region requires the introduction of a new source of economic stimulation, plus the development of reciprocal patterns of support between this source and the outlying hamlets and villages within its range of influence. Power parks offer themselves as potential sources of economic stimulation. And a body of economic doctrine known as "growth pole theory" suggests means by which reciprocal patterns of support may be fostered.[17]

Growth Pole Theory and Regional Development

When a settlement serves not only as a supplier of products to the citizens in its immediate service area but also as a stimulator of induced economic activity throughout an outlying region, it is called a *growth pole*. The main urban centers in most regional hierarchies function as growth poles. Prime cities must serve as the centers of capital formation for entire regions—and not only of financial capital but of intellectual and human capital as well, providing a diverse array of skills made possible by the city's broad educational opportunities and made available through its active labor markets.

Modern growth pole theory originated in the mid-1950s with the work of the French economist Francois Perroux.[18] France, an agricultural nation, had lagged behind England, Germany, Italy, and the Low Countries in per capita industrial output. Some suspected that excessive centralization of decision-making power in Paris had impeded the development of outlying districts. Thus Perroux's analysis of the dynamics of economic growth responded to a new interest in industrialization and regionalization as *linked, complementary* phenomena.

Within a given region, Perroux argued, a firm may gain a dominant economic position as a result of its superior negotiating strength (often a euphemism for its monopolistic or monopsonistic trading power) or from the economically strategic nature of its activity. Perroux held that the dominant firm should be large, capital-intensive, and enjoy a privileged position as a large-volume buyer of materials and seller of products. Such a firm can exert a degree of control over the activities of its suppliers and its patrons in the region disproportionate to its own cash flow. Therefore, control over the firm's own policies equates to leverage over activity throughout the region.

The notion of using power parks as stimuli to the emergence of regional cities—cities of the second sun—hinges on the potential of nuclear centers to serve as growth poles in Perroux's sense. Critical to this approach is the ability of a power park to create a long-term market for the output of a special regional reactor factor. Let us briefly consider why a reactor factory of the sort described in Chapter 5 would assume the role of a dominant or "propellant" firm.

A reactor factory, producing modular components for on-site assembly, would represent a basic departure from the typical procedure used in the manufacture of a commercial product. In the normal production cycle, materials are gathered from mines or quarries. These raw materials then move through the successive stages of milling and fabrication. They also move from one work site to another. Ore from the mines of Minnesota may become a pig of iron in an Indiana mill, then move to a plant in Ohio for rough shaping, and then to a New Jersey factory for finishing. At each stage in this process, the contribution of workers may be approximated by the economic value-added in the particular phase of processing. "Value-added" refers to the gain in market price of the product achieved at each new site as a consequence of the additional work there contributed.

With on-site fabrication and assembly, a much larger percent of the total value-added in nuclear plants worth tens of billions of dollars would be concentrated in the region of the power park. The reactor factory, rather than the power park itself, would function as the propellant by converting a national process into one largely concentrated in the region. So it would be the region that would bear the main impacts of the development—positive impacts in the form of new industries and hence new jobs, potentially negative impacts in the form of local inflation, shanty towns, and environmental threats. In Chapter 7 we shall consider the process of regional planning as a means of minimizing these negative impacts.

Summary: Power Parks and the Problem of the Dual Society

America's intermetropolitan frontier corresponds to the entire area beyond easy commuting reach of any of the urban centers projected to exist in the year 2000. The economic and demographic erosion of America's intermetropolitan frontier can be reversed only if the innovative skills of America's urban civilization are turned to the task of revivifying the hinterland. Such a reversal would define a worthy goal of national social policy. More and more Americans seem willing, if offered the economic opportunity, to move from the megalopolis to smaller cities and towns on the frontier.

The first requisite of a policy to stimulate the declining frontier is a plan for creating economic activity within regions suffering from shrinking opportunity. The concentration of generating capacity in power parks could support the development of special reactor factories. In their turn, these factories could concentrate economic activities that otherwise would remain fragmented and dispersed throughout the national supply network. Such a concentration answers to the prime criterion of a regional growth pole. The increment of steady employment—measured by the higher proportion of value-added that would occur within the region—might be expected to influence the entire hierarchy of towns, villages, and hamlets. The power park, with its associated reactor factory, would thus function as a propellant firm for the region or as an analogue to a prime city in the hierarchy of more highly urbanized regions.

Yet it should be recognized that growth pole theory, however attractive to the academic planner, raises issues of deep concern to practical politicians—in other words, to all who would have to face the difficulties of application. Nuclear centers *could*, as some opponents of power parks have contended, disrupt America's poorest, most sparsely settled rural regions.[19] And clearly, a power park could overwhelm any single small town, were such a community made to absorb the entire population influx and economic impact. The real question is: By what means might citizens in potential host regions try to ensure that planning would indeed proceed in a responsible, foresighted manner? Fortunately, mechanisms are available to incorporate social objectives—and specifically, the interests of citizens in a host region—as prime goals of a national strategy for sufficiency of electric supply rather than as mere afterthoughts of technical planning decisions.

Revision of the environmental review process, an achievement of ecology advocates in the 1960s, to emphasize the socioeconomic needs of frontier regions, suggests the direction in which to move. Such a move would effectively duplicate the process whereby environmentalists gained access to decisionmaking procedures from which they had, prior to passage of the National Environmental Protection Act, been largely excluded.

Notes

1. The studies of potential sites at Camp Gruber, Oklahoma and at Glasgow Air Force Base, Montana have been published by the U.S. Federal Energy Agency under the codes, FEA/G-75/384 and FEA/G-75/418, respectively. See also a study by the Energy Policy Analysis Group, Brookhaven National Laboratory of a possible Ocean County, New Jersey site for a 40-unit nuclear center: "Preliminary Assessment of a Hypothetical Nuclear Energy Center in New Jersey" (BNL-20594, November 1975); and "Nuclear Energy Center, Upper St. Lawrence Region" by researchers at Argonne National Laboratory (October 1975), a contribution to the NRC NECSS-75.

2. General information in Grace Lichtenstein, "Kaiparowits . . . ," *New York Times* (November 23, 1975), E3 and "Big Power Plan Dropped in Utah," *New York Times* (April 15, 1976), 1, 21.

3. See *North Central Power Study* (Billings, Montana: U.S. Bureau of Reclamation, October 1971), pp. 2-9, "Report on Phase I."

4. See the two AEC staff studies that have been combined into a single volume: U.S. Atomic Energy Commission "Evaluation of Nuclear Energy

Centers" (WASH-1288, January 1974). The first study considers the proposal for a nuclear park at the Hanford facility in Washington state; the second reports on an evaluation of the proposal for a large nuclear center at River Bend, north of Baton Rouge, Louisiana. See also C. Richard Schuller et al., "Citizens' Views about the Proposed Hartsville Nuclear Power Plant: A Preliminary Report of Potential Social Impacts," (Oak Ridge National Laboratory, ORNL-RUS-3).

5. Jim Kimball, "Aitkin resident weighs impact . . . ," *Minneapolis Tribune* (August 10, 1975), 1F.

6. E. Reis, E. Kasum, and J. Thorpe, "Energy Parks in Pennsylvania: Where, When and How," March 15, 1974, p. 11, available from any of the participating utilities; see also The Governor's Science Advisory Committee Panel on Criteria for Power Plant Siting, "Criteria for Power Plant Siting in Pennsylvania, A Discussion of Technological Considerations." An interesting account of the politics of the Pennsylvania power park proposal appears in the study "The Energy Park Experience in Pennsylvania" by Dr. Terry Ferrar, Director, Center for the Study of Environmental Policy, Pennsylvania State University (September 20, 1976), available in mimeo.

7. "Is It Time for a New Look at Solar Energy?" 27 *Bulletin of the Atomic Scientists* (October 1971), 32.

8. (New York: Oxford, 1959), Chap. 1.

9. The demographically oriented data in this chapter derive mainly from the report of the U.S. Commission on Population Growth and the American Future (John D. Rockefeller III, Chairman; Charles Westoff, Executive Director); see *Population and the American Future*, an Overall Summary of the Commission's analysis and findings, including illustrations and references; see especially p. 120.

10. Ibid., p. 36.

11. See generally Brian Berry, *Growth Centers and their Potentials in the Upper Great Lakes Region* (Upper Great Lakes Regional Commission, May 1969), pp. 4-5.

12. Figures taken from U.S. Census reports as summarized in "Rural Areas' Population Gain Now Outpolling Urban Regions," *New York Times* (May 18, 1975), 1, especially p. 44; and "The Population Shifting Goes On and On," *New York Times* (December 14, 1975), 5E.

13. Minnesota State Planning Agency, *Perspective on Land Use–1974* (University of Minnesota Center for Urban & Regional Affairs, October 1974), pp. 40-41. See also Neil Gustafson, "Region's Growth Stil Urban," *Minneapolis Tribune* (July 27, 1975), 13A.

14. For a moving and evocative discussion of the decline on the American intermetropolitan frontier, see Harry Caudill's *Night Comes to the Cumberlands* (Boston: Little, Brown and Co., 1973); and *My Land is Dying* (New York: E.P.

Dutton, 1971); to similar effect, in *The People Left Behind*, Report of the President's Commission on Rural Poverty, Washington, D.C., U.S.G.P.O., 1967.

15. Figures from U.S. Census data reported in ". . . Urban Crisis of 1960s . . . ," *New York Times* (March 23, 1975), 1, 46.

16. On the structure of the urban hierarchy, see Edwin Mills, "Market Choices and Optimum City Size" (with David deFarranti), 61 *American Economic Review* (May 1971), 340-345; "Welfare Aspects of National Policy toward City Sizes," 9 *Urban Studies* (February 1972), 117-124; and pertinent chapters in *Studies in the Structure of the Urban Economy* (New York: Resources for the Future, 1972).

17. See Morgan Thomas, "Growth Pole Theory: An Examination of Some of its Basic Concepts," in Niles N. Hansen, ed., *Growth Centers in Regional Economic Development* (New York: Free Press, 1972), 58. I have also drawn extensively from the survey volume edited by Antony Kuklinksi, *Growth Poles and Growth Centers in Regional Planning* (The Hague: Mouton, 1972), prepared under the auspices of the U.N. Research Institute for Social Development. The following two papers were particularly useful: Niles Hanson, "Criteria for a Growth Center Policy," and Morgan Thomas, "The Regional Problem: Structural Change and Growth Pole Theory."

18. "Note sur la notion de l'pole de croissance," in *Regional Economics*, David McKee, Robert Dean, and William Leahy, eds. (New York: Free Press, 1970), 94.

19. A study by Sam Carnes and Paul Friesema, "Urbanization and the Northern Great Plains" (Center for Urban Affairs, Northwestern University, May 1975), is typical of the work by scholars who rightfully fear the degradative social effects of major energy-related developments in America's rural regions.

7 Planning for the "Interregional Nuclear Bargain"

The idea of using nuclear centers as growth poles to stimulate the intermetropolitan frontier—can it work? Surely the idea *can* work, at least to modest effect in carefully selected regions of rural America. But whether the idea *will* work depends on a reversal of the traditional American approach to major construction projects in remote areas of the land.

The history of resource exploitation in the United States suggests that power parks could serve as degradative and disruptive intrusions, rather than as tools for the planned development of balanced, stable communities. The restrictions on degradative practices that have been enacted in environmental statutes would protect some areas against the worst forms of exploitation by would-be power park developers. Yet in order to ensure realization of the full potential of the growth pole concept, it will be necessary to move beyond the essentially negative, prohibitory achievements of protective environmentalism.

From Protective to Constructive Environmentalism

From the early Western mining towns, through the rape of Appalachia by lumber and coal interests, up to the lessons of the new strip mines near Gillette, Wyoming and Black Mesa, Arizona, the record of rapid economic development in America's intermetropolitan frontier has been one of social irresponsibility followed by severe community dislocation. Many rural areas of the United States have neither traditions of active civic involvement by highly educated electorates nor access to sophisticated counsel by experts who can contend with Big Business, Big Money, and Big Plans coming into the region in carpetbags.

Two past cases of obtrusive developmental efforts—TVA and the Alaskan pipeline—suggest the destructive as well as the constructive potential of power parks.

During the 1960s, the Tennessee Valley Authority—itself a kind of multistate power park—fell under attack by environmentalists who opposed the accelerated strip mining of Appalachian coal to fuel TVA's voracious fossil-fired electric plants. Yet over the longer reach of its existence, TVA has exerted an overwhelmingly favorable impact on the socioeconomic development of the once backward region. Its record stands as a proud achievement, not only in natural resource management but in human resource development as well. Since a single nuclear park could easily have about the same generating capacity as the entire TVA, the establishment of cities of the second sun would effectively require duplication of the Authority's capital plant dozens of times over across the land. But in each instance, the new construction would take a more concentrated form and would occur at locations preselected for their potential to support planned economic development.

More recent–and less encouraging–is the example of the Alaskan pipeline.[1] One of the most expensive private undertakings in history ($10 billion), the pipeline will bring crude oil from Prudhoe Bay on the Arctic Ocean to the ice-free port of Valdez on the Pacific, where it will be transferred by tanker to refineries in Washington state and California.

To safeguard Alaskan wildlife and the tundra wilderness, environmentalists forced major changes in the initial pipeline plan. In the worlds of Winthrop Griffith, a journalist who surveyed the construction scene near Fairbanks in mid-1975, "The caribou and the grizzly bears, the gophers and the Arctic lemming, the delicate tundra grasses and the forest of spruce around Fairbanks and south will survive."[2] But will Alaska's community fabric survive–survive the construction crews, the camp followers, the corruption of money? In their solicitude for endangered nonhuman species, pipeline foes erected few protective barriers against construction crews' threats to the social life of communities.

The social fabric of rural Alaska has proved every bit as fragile as is the ecology of the tundra biome. The pipeline marks a trail of galloping inflation and social dislocation as long as the project itself and reaching miles to either side of the right-of-way. A running boom town has disrupted local communities and fractured Eskimo tribal and village relationships. A transient workforce taking in too much whiskey and putting out too much money overloaded local recreational, service, and health facilities. Severe medical problems range from higher venereal disease rates among native Alaskans to alarming levels of traumatic injury among the workers.

The pipeline, then, promises none of the long-term socioeconomic benefits associated with the history of the Tennessee Valley. In fact, quite the opposite. But presumably, the most severely adverse social consequences of the pipeline might have been forestalled, just as most of the adverse environmental impacts of the project were anticipated and in the main avoided, if planning had been informed by the kind of social awareness that prevailed among the early visionaries of TVA.[3] At minimum, advocates of social protection might have given the pipeline the kind of concerned oversight that environmentalists provided with respect to ecological issues.

Inevitably, projects of the scale of TVA or the Alaskan pipeline have effects that are not routinely reflected in the dollar calculations of their costs and benefits. These effects are generally equated to environmental externalities. But there also exist other kinds of externalities–spillover effects in addition to air pollution, land despoliation, and aesthetic degradation. Indeed, "the environment" itself represents far more than those wilderness reserves which indisputably need protection from developers and industrialists.

The environment includes all the fabricated as well as the natural features of the human milieu. It includes too the nonphysical structure of social and economic opportunities that define an individual's "life-space." And the spill-over impacts of any major project, either for good or for ill, on the quality of day-to-day community life deserve attention at least as serious, and certainly as aggressive, as environmentalists have won for ecological considerations. A goal implicit in the concept of the city of the second sun is that of internalizing an awareness of social externalities in the planning process in order to reduce the incidence of negative effects such as have marred the Alaskan pipeline. The aim of *constructive* environmentalism—in contrast with the ecology advocates' traditional emphasis on *protective* environmentalism—would be to develop people's surroundings as needed to improve the economic and social, along with the natural, features of the environment.

Constructive environmentalism, then, would aim at extending certain gains registered by ecology advocates since the mid-1960s to the full range of externalities, social as well as environmental, that define the milieu of the American future. The specific objectives are: (1) to anticipate potentially helpful social spillover effects and (2) to provide institutional mechanisms whereby such effects may be encouraged to occur on a controlled schedule.

The years 1965 to 1975 saw the rise of environmental politics, with passage of the National Environmental Protection Act. May the decade of 1975 to 1985 mark the achievement of no less vigorous an effort toward social planning.

Extending the Gains of Protective Environmentalism

Section 101 of NEPA requires that any proposal for major construction projects be consistent with the goal of protective environmentalism—that of "preservation and enhancement" of the natural environment.[4] Intellectual leaders of the movement that culminated in NEPA, such as "the father of the environmental impact statement," Lynton Keith Caldwell of Indiana University, all along contended that narrow ecological goals only partially represent the range of issues in which the American public has an interest.[5] However, still underway is the search for administrative procedures whereby the protective measures embodied in NEPA can be used simultaneously to advance constructive social goals. Ironically, organized environmentalists have sometimes appeared as obstacles to the achievement of such social goals, even though the needed new procedures would easily evolve from the very provisions of NEPA that were aimed at securing the values of protective environmentalism. Most specifically in point are those provisions which require proponents of new projects to file "detailed environmental statements."

A Social Supplement to the Impact Report

Under Section 102 of the Act and pursuant to implementing regulations published by the Environmental Protection Agency and the Nuclear Regulatory

Commission, each utility must prepare an Environmental Impact Report describing the likely effects of a proposed new nuclear generating installation. Although the implementing regulations of Section 102 do evidence concern over social, demographic, and economic effects,[6] the transcendent preoccupation is with ecological issues–scarcely an inappropriate concern, given the origin of the Act.[7]

NEPA itself requires that environmental evaluations contain detailed data on:

1. The environmental impact of the proposed action.
2. Any adverse environmental effects which cannot be avoided should the proposal be implemented.
3. Alternatives to the proposed action.
4. The relationship between local short-term uses of man's environment and the maintenance and enhancement of long-term productivity.
5. Any irreversible and irretrievable commitments of resources which would be involved in the proposed action should it be implemented.

After receiving comments by anyone who chooses to offer an opinion on the proposed new facility, Nuclear Regulatory Commission staff members prepare a final Environmental Impact Statement for submission to the President's Council on Environmental Quality.

These procedures allow for representation of the principal private parties interested in the proceeding. These parties include, most prominently, the utility company (or consortium of utilities, or perhaps an investor-owned G-T of the kind described in Chapter 5) seeking certification of the new power station. Also, property owners in the neighborhood of the plant site sometimes enter the licensing process to oppose the proposed development.

In theory, any member of the public may review the utility's evaluation of its proposed construction. In fact, comments from the public are mainly prepared by spokespersons for groups whose particular interests conflict with the siting plan under review. The review procedures under NEPA therefore do admit the viewpoints of adverse interveners. But such interveners press essentially negative, and sometimes frankly obstructive, interests–the aims of those aligned against any threat to the stability, wholesomeness, or aesthetic qualities of the environment.

The true public interest, however, may not lie between the positions of the pro's (the utility) and the anti's (the environmentalists). The public interest may lie elsewhere. In this event, it might fail to gain adequate representation in an adversary process structured to balance the private interests of developers and landowners against the negatively oriented interests of environmentalists and no-growth lobbyists. Positive public interests, such as regional economic stimulation and community development, have an imperative claim on the planning of

America's future. It should be possible to revise current licensing procedures in order to reflect the importance of that claim.

The planners of a proposed new nuclear plant must respond to an elaborate set of informational requirements as set forth in the Nuclear Regulatory Commission's 77-page Regulatory Guide 4.2.[8] The "Standard Format and Content of Environmental Reports" required by Guide 4.2 calls for design data on the proposed plant, reports on possible alternative sites, and estimates of environmental effects. These effects are to be discussed in the finest detail. In addition to such standard categories as ecology, geology, meteorology, and hydrology, some attention is directed to the impact of new construction on "regional historic, scenic, cultural, and natural features." The only other notable evidence of concern for nonphysical impacts occurs in the provision for a cost-benefit analysis of the proposed plant's economic and social effects–a single chapter out of the 10 substantive sections in the standard format.

A typical Environmental Impact Report for a 1000-megawatt nuclear plant runs to thousands of pages. But the value of the information can rarely be measured by its quantity or even by its precision, divorced of the specific purposes for which the data were gathered. Information must show a suitable degree of particularization to the specific problem on which the data will be brought to bear.

Since the categories of information required by Guide 4.2 are standardized, much of the reported data are in the nature of boilerplate verbiage. (Reportedly, the U.S. Army Corps of Engineers, facing an annual requirement under NEPA to complete dozens of environmental reports, developed mimeographed inserts covering items of common applicability for inclusion in its Impact Statements.) Standardized reports can leave much to the discretion of the respondents. Inevitably, they are "decontextualized" in the sense that the format of Regulatory Guide 4.2 applies to every new reactor, regardless of where it may be sited.

Standardized requirements can effectively guide analysis when the ultimate objective is protective. A general exclusion of behavior likely to produce the prohibited consequences, regardless of when or where, may then suffice. But as the goal shifts from protective to constructive environmentalism, positive objectives often cannot be achieved, or even meaningfully discussed, without reference to specific contexts. Economic growth defines such an objective, and specific portions of the intermetropolitan frontier define such contexts.

Modest additional informational requirements–no more than marginal increments to NEPA's already considerable demands on petitioners for new power plant licenses–could help redress the present overbalanced concern for ecological issues. Needed is a fairer appreciation of regional objectives: for example, changing the mix of townspeople and rural dwellers, or altering the relative rates of growth in particular economic sectors within a region.

A detailed "social supplement" to the Impact Report for a proposed power plant, as is already the practice for water resource and interstate highway projects, would explicitly put demographic and socioeconomic goals on a par with traditional ecological values. In these supplements, utilities might be required to indicate how power park construction, as compared with development in the dispersed siting mode, would help to achieve positively enunciated regional social goals.[9]

The Social Supplement and Regional Planning

The requirement to complete a social supplement would impose an added burden of proof on any G-T desiring to expand generating capacity within a region. Such a G-T would have to show how the kind and amount of proposed new construction might contribute to specific demographic and socioeconomic goals. Acceptability merely from the standpoint of more narrowly defined "preservation and enhancement" requirements would no longer suffice.

Of course, any effort to particularize environmental review procedures—and especially to particularize them along regional lines—would have dangers. Certain advantages of standardized, generic categories might be lost.

The generic categories of Guide 4.2 reflect more or less objective judgments by expert drafters of NEPA's implementing regulations. Such judgments reflect no regional biases. They do not skew the respondent's analysis toward an excessive concern with land acquisition over cooling water availability, or vice versa. The opposite might occur if either an Eastern or a Western regional orientation prevailed. Moreover, nationally standardized impact standards are fixed and open. They protect the public against uncertainty. They also help to prevent arbitrariness in the behavior of officials who will review a utility's submission.

By contrast, particularization of impact report requirements along regional lines would imply the substitution of a series of judgments about socioeconomic potentialities and problems in specific regions. Informational requirements could be rewritten as many times over as there were regions. But who would choose the regional borderlines? Who would decide when a given regional plan was satisfied or needed revision? Leaving such value-laden judgments wholly to the bureaucrats who write federal regulations or to the proponents of new projects—utility planners, or more generally, members of the nuclear community—could hardly represent much of an improvement in planning procedures.

The principle suggested by this analysis is as follows. When community leaders decide for power park development, they should have the right to devise a regional plan and then to invite development *on conditions specified in this plan.* It seems unlikely that excessive numbers of regions would be involved: regionalism is *not* a formula for anarchy or for the proliferation of nuclear centers.

The response of spokespersons in most regions to proposals for parks might be "No thanks." The deterrent to a proliferation of regional authorities, each specifying particularized socioeconomic objectives, would lie in the potential attractiveness of a nuclear center *only* where the benefits of economic stimulation might overbalance the possible environmental disadvantages. As discussed in Chapter 6, on the intermetropolitan frontier—and perhaps there alone—recent demographic and economic trends might combine to create incentives able to overcome the deterrent.

Lest a nuclear center be "rammed down the throat" of a hostile community, the members of such a community, after deciding that any positive socioeconomic gains possible with a well-planned power park would be outweighed by environmental disadvantages, could simply elect not to set up the regional structure. For want of the necessary mechanisms to bring a new facility into the area, would-be developers of a power park would have to turn elsewhere—to deal with officials in a region where the balance of advantages and disadvantages is perceived differently.

The regional authorities could take different forms in different areas of the country. In some regions the preferred form might be a governmental structure, perhaps on the model of existing interstate river basin commissions. Or the regional authorities could be chartered as multicounty development corporations working in close coordination with local officials on land use and zoning policy, road and school planning, and general intergovernmental cooperation to ensure the realization of community development objectives. These development corporations could be chartered by the U.S. Commerce Department's Economic Development Administration and might enjoy some preferential access to federal funds. But their actual memberships would be locally constituted to represent the needs of the immediate multicounty impact area.

The difficulties in establishing regional authorities cannot be understated. Political theorists have long debated the requisites of an acceptably "representative" set of institutions. In many of the medium-sized cities of America—for example, as the urban historian Daniel Elazar has shown, in the "cities of the prairie" throughout the Midwest—oligarchies consisting mostly of local business leaders have assumed control over the destinies of whole communities.[10] When confronted with the prospect of massive infusions of capital into a given area, business-oriented booster groups might tend to emerge in order to capture the economic benefits of power park development—not for their regions but for themselves. Emergent groupings of this sort would presumably not qualify as legitimately representative bodies whose acts might be regarded as expressing the will of the entire population in a multicounty impact area.

Congress would probably have to specify acceptable procedures for the constitution of a body before it would be accepted as authorized to speak for the citizens of a region. It is heartening to note that since the mid-1930s, the U.S. Supreme Court has frequently pronounced on the requirements of acceptable democratic procedures. Thus a body of doctrine exists that might be applied to the setting up of regional authorities.[11] Considerable latitude remains in the interpretation of such Supreme Court rules as "one man, one vote." Any set of constituent procedures that reasonably falls within this area of latitude would presumably be found acceptable as a mechanism for choosing of the members of the regional authority.

The Interregional Nuclear Bargain

A regional planner is not a designer of utopias. His is rather a very practical task–to anticipate those critical policy choices which can accelerate or retard certain developments in the natural "lifecycle" of a region as it emerges from stagnation to self-sustaining economic growth. The economic theory of regional development describes that natural lifecycle. The regional plan would apply the theory of regional development in a chosen multicounty impact area.

What might a regional plan look like in practice? The plan, prepared by members of the regional authority, would formally state the goals of constructive environmentalism as they apply to the multicounty area. It would define the particularized information required of would-be developers in the social supplement of their Environmental Impact Report. Authority to redraft the plan as new problems are uncovered and new opportunities perceived–subject to any prior contractual obligations incurred between the regional authority and the developers–could ensure a fair measure of continuing control over the course of the nuclear center.

In a typical case, an Impact Report would be prepared by an appropriate regional G-T in response to the NEPA requirement for a "generic" environmental study as the initial plans for a nuclear center are being readied. Subsequently, the G-T would file addenda to the additional unit-by-unit Environmental Reports that must be sent to the Nuclear Regulatory Commission each time an additional doublet or quad of reactors is to be put into production. In these addenda, the G-T would be required to indicate the likely incremental impact of new units on the goals enunciated in the constantly updated regional plan.

Under this conception, the regional plan should represent an offer to American society at large–an offer to bear some of the environmental burdens of providing a share of the nation's electric power in exchange for long-term development opportunities and the right to exercise reasonable control over the course of nuclear planning. And if the regional plan corresponds to an offer, preparation of a responsive social supplement by a developer would correspond to an acceptance. The regional plan thus becomes a kind of prelude to an "interregional nuclear bargain" whereby the region gains the means to recover from economic stagnation and whereby populations in more distant areas gain assurance that sites will be available for the reactors needed to supply their electricity.

The Export Effect

Because the nuclear complexes of the 1990s would provide suitable locales for the introduction of fusion torches and eventually of the fusion electric plants of the twenty-first century, a properly framed regional plan should accommodate not merely the initial power park but should also commit the locale to long-term development as a true city of the second sun. Such a commitment would identify the region as a basic power supply area, exporting electricity to the rest of the nation in much the pattern whereby the Gulf Coast states export natural gas to the North, or whereby the Plains states export grains to the East and West.

Given such a conception of the regional plan, the long-term significance of a regional commitment to power park development becomes clearer in light of the projection in Figure 7-1 of U.S. nuclear demand through the early decades of the twenty-first century. We have seen that the coarsened national grid implicit in a policy favoring nuclear centers would require the selection of approximately 27 main power exporting regions, most of them on the intermetropolitan frontier. If each power park initially averaged fifteen 1,200-megawatt light-water reactors, then some two-thirds of the nation's projected nuclear demand in the year 2000 could be met from these 27 centers.

Thereafter, electric demand should continue to grow. Growth in such demand will stimulate each electric power supply region by means of the second category of effects noted in Chapter 6, the export effect. Just as growing demand for natural gas causes capital to flow into the Gulf Coast states so methane can be recovered for export to Northern cities, growing electric demand should lead investors to continue installing generating capacity in those regions which have the highest potential for further development as suppliers of power.

The nuclear centers likely to be under construction through the 1980s imply the possibility—indeed, the necessity—of developing whole communities of highly skilled construction workers, sophisticated managers, and nuclear-rated plant operators. A regional nuclear workforce would be ready-trained and available to man the plants needed to meet rising demand. Roughly two-thirds of the nation's projected electric demand in the year 2020 can be supplied from the same 27 power-exporting areas selected for the first nuclear centers. The number of units per park would grow from 15 to 40. Up to half the 40 reactors in a given park might consist of new fusion units.

This number—40 units—sets a critical threshold figure. As emphasized in Chapter 5, the economic case for clustering reactors derives from the savings to be achieved with more efficient production techniques. These gains in turn require the recruitment and maintenance of a skilled labor force. It follows that

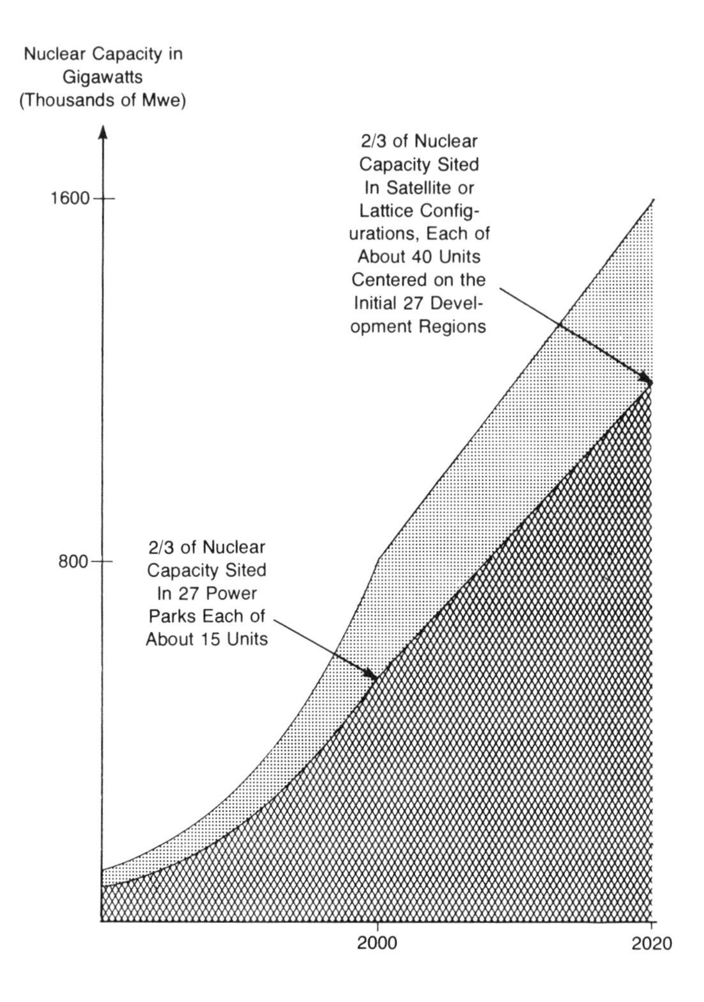

Figure 7-1. Nuclear Demand Through the Early Twenty-first Century.

the scale of forseen new development throughout the region must guarantee long-term employment for nuclear designers, construction workers, and managers. In order to underwrite such a guarantee, the equivalent of 40 light-water units would have to be built within the region on a schedule that ensures the start of a rebuilding cycle as soon as the first construction cycle ends. That is, the first nuclear unit will get decommissioned just about when the fortieth is completed. Hence the regional workforce, instead of disbanding, could simply turn to replacing the units from the initial construction cycle. Of course, the number 40 is subject to adjustment, since later units in the schedule, if based on fusion, could have more than twice the megawattage of LWR units built early in the series.

(What if reactors are brought on line so quickly that the first units will have years of useful life remaining when the last unit in the first cycle has been built? Or what if plans for ultimate construction in the regional plan fall short of the 40-unit equivalent critical scale? When the pace or scale of envisioned reactor construction cannot ensure permanent employment, a requirement emerges for additional industrialization in order to support a levelized workforce even after the power park has been completed. Chapter 8 includes a discussion of strategies for meeting this requirement.)

As the number of nuclear units in a region rises from 15 to 40, the deployment of generating capacity would probably increasingly take on the aspect of satellite or lattice siting. That is, an initial 15-unit site would, once completed, begin to function as a staging base for satellite sites, so the additional 25 or so units need not necessarily be brought within a common fenceline.

Exactly how the total envisioned 40 units would be deployed around the region may be left to technical evaluation of the needs and the potential efficiencies of each unit. But the process of drafting (and periodically revising) the regional plan would represent much more than an exercise in technical appraisal of gains and costs. It would represent a continuous testing of communities' values, leading in some cases to a long-term political commitment to absorb certain of the risks–even as it profits from the benefits–of becoming a regional city of the second sun.

Summary: Extending the Gains of Protective Environmentalism

The long time frames involved in power park development, the size of the investment, and the importance of stable engineering, construction, and managerial cadres suggest a need to hedge the risks of all participants well out into time. An explicit schedule of objectives (hence the regional plan) backed by the good faith of responsible authorities (hence the need for extreme care in constituting the regional authority) seems a constructive, if by no means foolproof, means toward that objective. Crucial questions–Who shall decide the boundary of a power-exporting region? Who shall decide compliance with the regional plan?–must yield to political solution, not to simple technical answers, for ultimately, political representation and accountability must be relied upon to ensure that those who invite development *do* do so on behalf of the citizens in the region.

From the standpoint of citizens in the locale, the framing of a regional plan would represent the adoption of constructive environmentalism as a principle of action. To gain the socioeconomic benefits of an interregional nuclear bargain, certain values of protective environmentalism might have to be foregone. The development of a region as a major national energy supply area implies a population buildup. It implies the laying of new roadways and the condemnation of transmission corridors. It implies a continuing effort to manage the water resources against the cooling requirements—and potential thermal effects—of the nuclear plants. Most important, it implies a local radiation hazard of small but undeniably real proportions.

Whether these burdens should be accepted or not should be decided by those who would mainly have to bear them. A regional framework for energy planning represents an attempt to devise institutions that can make such choice possible. The aim would *not* be to reverse the gains already registered under the rubric of protective environmentalism, but to give constructive environmentalism its due place in the hierarchy of social values.

Notes

1. See my *Energy, Ecology, Economy*, at pp. 106-107, 111. I have also taken profit from the history of the Alaska pipeline project by Richard Horrigan, prepared under the auspices of the American Enterprise Institute, made available to me by Dr. Edward Mitchell, Director of the Institute's Energy Project.

2. "Blood, Toil, Tears and Oil," *New York Times Magazine* (July 27, 1975), p. 41—also a helpful article in its account of the social impact of the pipeline.

3. Arthur Morgan's autobiography, *The Making of the T.V.A.* (Buffalo: Prometheus Books, 1974), especially Chap. 12, provides illuminating insight into the early "social theory" behind the Tennessee Valley Authority.

4. 42 U.S.C. 4321 *et seq.*

5. In private communications, my friend Professor Richard Andrews of the University of Michigan called my special attention to the work of Lynton Keith Caldwell on the original philosophy of the Environmental Impact Statement.

6. On the background and applications of the NEPA, see Frederick R. Anderson, *N.E.P.A. in the Courts* (Baltimore: Johns Hopkins University Press, 1973, for Resources for the Future), especially Chap. VI.

7. See the "Symposium issue on Environmental Impact Statements," 16 *National Resources Journal* (April 1976), especially H. Paul Friesema and Paul J. Culhane, "Social Impacts, Politics and the Environmental Impact Statement Process."

8. "Preparation of Environmental Reports for Nuclear Power Stations" (January 1975).

9. The emphasis on regional regulatory approaches draws heavily on the ideas of Arvin E. Upton and E. David Doane, of the National Academy of Public Administration Panel on Energy Parks. See especially Doan's statement to the NRC Office of Special Studies, Public Meeting on Nuclear Energy Center Site Survey, June 17, 1975. See also The Mitre Corporation Report for the NECSS, "Nuclear Energy Center Workshop East," (NTR-6967), especially the report of Panel III, pp. 52ff. In addition, I have profited by my association with the Regional Energy Studies Program of the Brookhaven National Laboratory, and from the conferences and papers of that program—particularly Philip Palmedo, "Regional Energy Analysis: A State and National Need" (BNL-19466, November 25, 1974). See also pp. 94-98 of Berlin, Cicchetti, and Gillen's *Perspective on Power*, and the now standard descriptive account of American experiments in regional government, Martha Derthick, *Between State and Nation* (Washington, D.C.: The Brookings Institution, 1974), *passim.*

10. *The Cities of the Prairie* (New York: Basic Books, 1970), especially pp. 208ff.

11. Although the courts once avoided cases dealing with allocations of political power, a body of relevant doctrine began to emerge with such cases as *Baker v. Carr*, 369 U.S. 186 (1962) and *Reynolds v. Sims*, 377 U.S. 533 (1964).

8 Developing the City of the Second Sun

Three developmental phases may be identified in the natural life cycle of an emerging city of the second sun. First: a period of site preparation, followed by actual nuclear center construction, and then operation of the first generating units (while further construction proceeds). During the early years of this initial phase, the complex need consist only of those facilities which are necessary for reactor fabrication, for on-site assembly, and (in the case of a park with co-located fuel cycle facilities) for plutonium reprocessing and fuel fabrication to close the nuclear cycle. Crucial is the need to assemble a permanent workforce with the skills needed to construct the park. Workers should reside within ready commuting distance of *all* envisioned units, whether they are to be sited in the aggregative, the satellite, or the lattice mode.

Within this initial phase, the regional plan may provide for specified "steps" corresponding to different scales of development. The first step might correspond to a minimal regional commitment to development of a 15-unit fission-based power park, but without further industrial development of any sort. The second step could equate to a somewhat more ambitious development objective—say, to bring construction to the level of a 25-unit nuclear center with co-located fuel cycle facilities. The drafters of the regional plan would identify a whole sequence of possible additional steps, right up to the 40-unit mixed fission and fusion regional energy park, together with a full-blown effort to develop heavy industry in the region.

At some point in such a sequence, the basic demographic characteristics of the region would change. A once rural area profiting from the tax receipts of a small nuclear center would begin to show the characteristics of a true city—a city of the second sun.

After the initial construction phase comes the phase of urban growth, characterized by the appearance of a more extensive physical and sociopolitical infrastructure to support a growing list of services for citizens in the region, *plus* certain recreational and cultural amenities to meet the needs of a growing population, *plus* some industrial development beyond that immediately associated with electric power generation.

In the third and culminating phase of regional growth, a growing proportion of needed goods may begin to be produced within the immediate vicinity of the power park, instead of being imported from distant manufacturing centers. Jane Jacobs has termed this process the *explosion into cityhood.*[1] Explosive economic development could underwrite the maturing of regional cities of the second sun.

Power Parks and Total Energy Communities

As discussed in the thumbnail history of American urban development presented in Chapter 6, the nation's leading cities grew only when people turned "free goods" from nature–the tillable hinterland of a St. Louis or a Minneapolis, the mineral deposits near a Pittsburgh or a Dallas, the harbor of a New York or a San Francisco–to productive account. But the geographical endowments that were to prove decisive for the American cities of the past were givens, not variables. The founders of past settlements could not *choose* the locations of arable acreage, mineral deposits, or natural harbors. By contrast, planners of cities of the second sun can by careful choice of the sites of nuclear centers select the locations of the dominating physical influences of some of America's new population concentrations. A power park could confer much the same kind of locational advantage that favorable local harbor conditions or nearby oil deposits formerly implied.

Like a natural geyser field, a nuclear center may be viewed as a source of stored heat, potentially available to generate electricity. But excess heat from a nuclear reactor–the two-thirds now lost as a result of thermodynamic inefficiency–cannot easily be tapped for productive use. Hence as emphasized in Chapter 5, heat dissipation presents one of the most serious challenges to American nuclear policy.

Might the quantities of waste heat implicit in a nuclear center be turned from an environmental drawback into a qualified virtue? The original proponents of agroindustrial complexes thought so. They saw the availability of heat for sale to desalination plants and nearby industrial firms as a leading attraction of power parks.

The Technology of "Heat Control"

The development of markets for the excess heat from nuclear centers would reduce the thermal load on the environment. Heat that warms a home, drives a synthetic chemical reaction, or speeds the growth of cash crops in a greenhouse need not thermally pollute the nearby waters. So if waste heat from power parks can be captured for use in socially valuable activities, then the burden imposed on the ecosystem will be eased.[2]

Waste heat from a power station will prove useless to a potential residential, industrial, or agricultural consumer unless it can be provided when and where the buyer needs it. Deliverability of heat implies a physical hookup between the power plant and the industrial facility. Because such hookups are difficult to retrofit, maximum recovery of heat through comprehensive design of entire energy-consuming systems–the so-called total energy concept–calls for a prospective strategy. That is, total energy has its primary value as a guide for the planners of facilities yet to be built. Nuclear centers with co-located agricultural complexes and new towns are of this description. Power parks may be designed with provision made in advance for the necessary physical connections between power plants and heat-using industrial firms or agricultural complexes.

Unfortunately, heat degrades rapidly with time and distance from its source. So with existing technology, unless the secondary consumer can be sited in extreme proximity to the power plant, the thermal effluent will have degraded during transport to a quality below the threshold of usefulness. Underground installation and improved insulation of pipes to carry waste heat from the power plant to the secondary consumer would permit more efficient use of the thermal effluent. And a relatively new technology—the heat pump—may go far to make total energy planning a realistic design concept, at least in certain applications, for future power site planners.

A heat pump may be thought of as an engine "working in reverse." A normal engine works between a hot source (the heat supplied by the combustion of oil, coal, or some other energetically charged substance) and a cool reservoir. A fraction of the thermal energy in the hot source is converted into mechanical energy to turn an axle or to push a load. The unconverted fraction is exhausted to the cold reservoir. A heat pump, which runs in just the opposite direction, is driven by an external motive power, such as electricity. The pump draws diffuse, low-grade heat *from* the cool reservoir and exhausts it *to* the hot source—viz., the space to be heated. With an engine, the hot source must be renewed by adding fresh fuel, while the cool reservoir is "thermally polluted" by exhausted heat. With a heat pump, the cool reservoir gets colder as heat is withdrawn, while the counterpart of the hot source receives the efflux of heat.[3]

The "thermally polluted" cooling water from nuclear plants—from which one would desire to withdraw heat—might be regarded as a potential cool reservoir for an entire community. In a total energy system the homes, the aquaculture ponds, and the greenhouses of the community into which one would desire to add heat become the counterparts of a conventional engine's heat source.

Heat pumps cost more to run than do resistance heaters, which throw off heat as current passes through a resistive conductor, as in an electric toaster. A resistance heater puts out a single Btu of heat for every Btu-equivalent of current passed through its coils. By contrast, a heat pump can be made to yield up to 3 Btus for each Btu used to pump low-grade heat from the "cool reservoir" up to the higher quality state. Given this already achievable level of efficiency, and given the generally increased prices expected for oil and gas in the 1980s and beyond, heat pumps of improved design seem likely to become more nearly competitive on a dollar basis with the older, less-efficient forms of space heating.

Used as "boosters" of low-grade heat in the pipes carrying thermal effluent from an energy park, heat pumps could serve both agricultural and residential customers, as well as the space heating needs of industrial firms (see Figure 8-1).

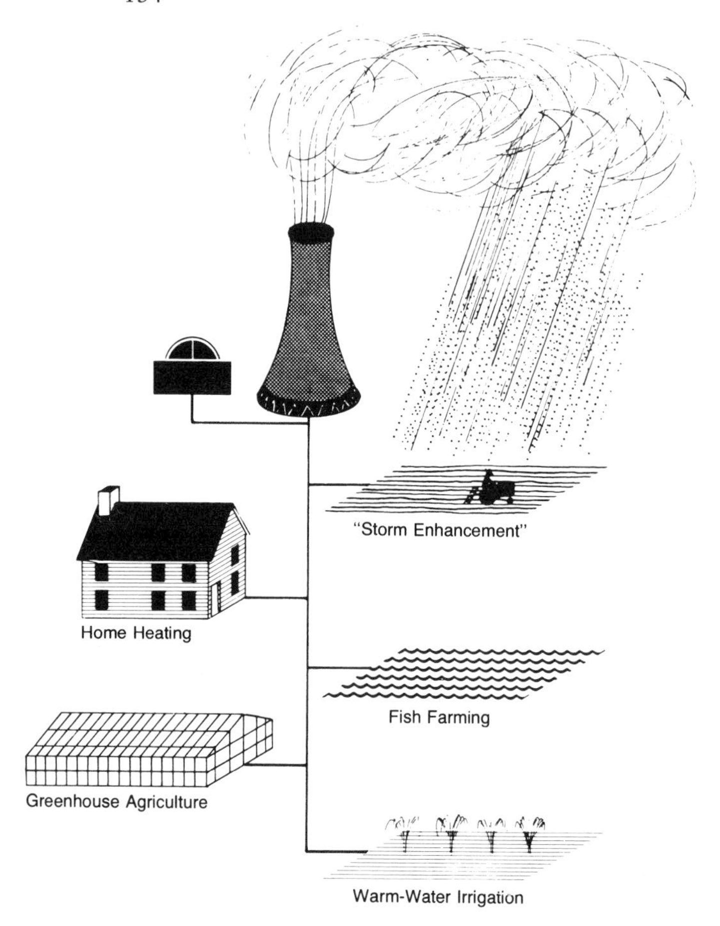

Figure 8-1. Residential and Agricultural Markets for Waste Heat.

Notes to Figure 8-1. In Eastern areas, the evaporate from cooling towers increases precipitation within the watershed from which makeup water is drawn, thereby accelerating the hydrologic cycle. In most cases, the augmented rainfall in the region would have a beneficial impact on crop raising.

Some of the thermal effluent from a power park could also be used, with the help of electrically driven heat pumps, for central heating of clustered homes.

Given appropriate advanced planning, waste heat from a nuclear center could serve, too, as an input to the local agricultural economy. If ponds or channels are used to supplement wet cooling towers, the heated water may be used in aquaculture, so long as it is kept within the specified thermal tolerances of the organisms under cultivation. The channels would then serve double duty–to dissipate the thermal effluent from power plants and also to provide a favorable environment for "fish farming" and for the raising of oysters and clams. Spray-cooling has also been shown to be an effective means of irrigating truck-gardens and orchards, especially in areas where thermal control is a factor in crop growing. Heated water as an irrigant, for example, can contribute to the prevention of frost-kill. Finally, heat pumps might be used to "reconcentrate" the dilute heat of power plant cooling water as a means of improving environmental controls within commercial greenhouses.

Careful moderation of the rate of thermal discharge to stock ponds can increase yields of high-protein food sources. Thus waste heat might be used to accelerate the growth of important fish species: catfish, carp, and sole; lobsters; oysters and clams. Controlled spraying of heated water from power plants has been shown to aid in open-field farming, as well as in the management of orchards. Waste heat can also be used to ensure optimum climate conditions in artificially closed agricultural environments, with phenomenal increases in yields. A single 1000-megawatt plant, it has been estimated, could maintain more than 200 acres of greenhouses at 80° Fahrenheit, given an average outside temperature of 40°.

The possibility of new housing construction as an integral element of power park development suggests planned centrally heated communities. Families could then use waste heat to warm their homes. (Of course, any routine tritium emissions would be diverted to special streams of coolant flows, so heated water for either agriculture or home heating would not be irradiated.)

Centrally heated communities have been proved feasible in Finland, Iceland, and especially in Sweden, where more than 27 such new towns have been built since 1951.[4] The heat pump suggests a means for realizing total energy even in communities with substantial dispersal of secondary consumers from the central heat source. Cities of the second sun would be of this description, since a radioactive exclusion area would intervene between the reactors in the park and any nearby population buildup. As heated cooling water moves through this exclusion area, it would degrade from "hot" to "warm" to "tepid." Heat pumps run on electricity from the park would then reconcentrate the degraded heat at distances up to several miles away from the point of effluence.

Computations by A.P. Fraas of Oak Ridge National Laboratory suggest that, at prices prevailing in the late 1960s, cooling water could be transported as much as 20 miles from a generating plant and still deliver economical heat to a load center.[5] The Fraas estimate makes no provision for the use of heat pumps. Allowing for such use, and assuming advanced planning to lay efficient underground steam pipes or hot water mains from reactors to an area of expected community development, the possibility of total energy communities located well beyond the most conservatively drawn fenceline about a nuclear center seems unquestionable.

The Industrial Base of a City of the Second Sun

The movement of industrial firms into the locale of a nuclear center, subject to the terms of the regional plan, would illustrate the third growth effect noted in Chapter 6, the agglomeration effect.

The Agglomeration Effect

The economic concept of agglomeration is central to an understanding of the dynamics whereby new industry might be drawn into the environ of a power park. Agglomerative industrialization refers to a twofold process. First some going operation produces certain externalities, such as waste heat, which offer themselves as free (or at least as relatively cheap) goods to potential consuming firms. Second, the potential consuming firms move into the vicinity in order to gain the benefit of the externalities. The availability of a highly skilled labor force or an emerging community of nuclear safety specialists would provide additional attractions to firms seeking new sites.

As industries gather to reduce the transfer costs of goods which they may purchase more cheaply by locating close to the source, the population increase broadens the market for goods and services. So the expanding economic activity underwrites the growth in consumer demand needed to attract retail firms. Employees of these retail firms further swell the market, adding another fillip to the growth process. The number of pathways whereby a given type of agglomerative economy will induce further stimulative effects increases with the growth of the region itself.

Because industrial development brings new population into a region, agglomeration takes on the character of an "intervening variable"–the variable which accounts for the difference in population between new residents who are initially drawn by jobs at the power park and the total number of additional people who finally take up residence in the region. Thus control over the pace and ultimate level of industrial development implies control over social and demographic development of the vicinity.

Initially, the most powerful agglomeration effects following the location of a power park (with its associated regional reactor factory) would occur at the level of industrial location and expansion. By co-locating near a nuclear center, certain firms could profit from the availability of ample process heat and reliable bulk power.

Extensive empirical studies of regional growth patterns are available to suggest a typical sequence by which the stimulative effects of a new industry are transmitted through the region. Consider a firm in a capital-intensive industry, such as electric generation and transmission. Such a firm will have decreasing long-run average costs. That is, as output expands from a plant already in being, the marginal cost of producing the next item should diminish. The average cost of all units sold can go down, since the capital investment in the plant may then be spread over a larger number of units produced. So as the propellant firm boosts production from a given plant, it may lower prices. These price reductions should translate into lowered costs for dependent firms. The latter too can expand output as their own ability to lower prices enables them to widen their markets.

Ideally, lower prices charged by the original stimulative firm effectively increase the income of all customers who directly or indirectly rely on that firm's output. This effective increment, *if intelligently spent within the region*, may provide the margin of new capital needed to underwrite a continuing spiral of growth.

The pressure of shortages within the region—shortages created as firms, anticipating a more favorable cost base, seek to expand their output with an eye to broadening their markets—spurs further growth. A kind of revolution of rising expectations further stimulates demand, further strengthens the regional market—thereby adding to the self-sustaining momentum of the spiral.

By locating next to a power park, an industrial firm may position itself to purchase heat or electricity more cheaply than either could be produced within the firm's own production lines. The single most obvious kind of co-located industry—nuclear reprocessing, fuel fabrication, and perhaps even uranium enrichment facilities—could itself absorb a significant fraction of the output from a nuclear center. Fully 15 percent of the electricity from the proposed 26,000-megawatt park at Hanford, Washington would be dedicated to supplying co-located fuel cycle facilities.

Industrial Users of Process Heat

The use of high-grade heat from a nuclear center—say, steam at temperatures high enough to drive an electric-generating turbine—imposes two requirements: (1) diversion of some heat away from power production; and (2) location of the using industrial firm close enough for thermally efficient transport of the steam.

Consider, for example, the use of high-grade heat to desalinate water in an agroindustrial complex. If some of the steam from a reactor gets diverted from power production to evaporate briny water, the plant will operate at a lower efficiency than can be attained by a power-only unit. Yet in compensation, an estimated 70 million gallons of desalted water per day might be produced for every percentage point of reduction in thermal efficiency.

Other potential users of process heat are petroleum refineries, related petrochemical complexes, and steel mills—firms that need steam at pressures of 200 pounds per square inch and higher.[6] Of course, a refinery can easily produce steam of this quality by burning its own feedstocks. But use of fossil reserves to raise process steam for stationary consumption represents an inferior use of the earth's depleting hydrocarbons. For this reason, the National Governors' Conference in 1975 recommended further study of nuclear-based power parks located close to the nation's coal fields (e.g., in the North Central Plains states), producing process steam to gasify coal—thus, in the words of the Conference report, "simultaneously solving the air pollution problems of sulfur emissions" and putting coal in a form for use as a home-heating fuel.[7]

The AEC's agroindustrial complex studies, from which the figure on desalination efficiencies come, were completed in the late 1960s and early 1970s, years of extravagant expectations for nuclear power.[8] Typically, for example, these studies assumed a produced price of electricity in the range of 3 mills per kilowatt-hour. The tenfold deterioration of nuclear costs since the late 1960s inevitably raises doubts about the economic practicality of schemes (such as designs for mammoth desalination plants) that may have seemed perfectly sensible to the industrial engineers of a few years ago.[9] Yet intensifying concerns over global resource shortfalls—and not least, over impending shortfalls in the availability of potable water—suggest that more modest variants of the AEC's initially envisioned agroindustrial systems may yet find economic justification in the regional development plans of the future.

Suppose, then, that a power park complex of the early twenty-first century includes a series of evaporative devices for flashing off potable water from brines, leaving saline residues behind. Networks of chemical interdependence can then be used to recycle the residues of this process as feedstocks into other product lines.[10] Compounds of chlorine and industrial caustics may be extracted from saline residues. Hydrogen can be recovered from water by electrolysis and then used as a fuel or else combined with nitrogen from the air to make ammonia-based fertilizer.

Studies of agroindustrial complexes have been conducted at Brookhaven National Laboratory. These studies suggest that agricultural wastes may also be combined with hydrogen and carbon dioxide extracted from sea water, from the air, or captured as a byproduct of steel mills in order to synthesize alcohols. Further chemical conversion processes show promise of producing gasoline substitutes ("synfuels") through the dehydration of these alcohols.[11] Proponents of synfuel complexes seek to find substitutes for oil, gas, and coal. Once burned, these fuels are gone forever. Petroleum and coal find their highest-valued uses not as fuels but as raw materials to be incorporated in plastics, in fertilizer, and in chemical compounds. Heat may be obtained from alternative sources, such as nuclear reactors, but the fossil feedstocks to the petrochemical plants of an industrial society cannot be easily or cheaply duplicated.

The preservation of nature's fossil stocks emerges as a direct counterpart to the "thousand-year problem" of radwaste disposal. Nuclear opponents rightly urge the necessity of a long time horizon when evaluating the radiological dangers of fission power. But a similarly extended time frame must also be applied to any evaluation of alternative plans for reliance on fossil-fired electricity. Known American reserves of coal can, it is true, sustain growth in energy consumption for decades. Yet eventually some generation of Americans will confront the receding edge of the continent's fossil fuel resources. Prudence suggests an effort to conserve existing stocks of hydrocarbon fuels or else to maintain them (by incorporation into synthetic materials) in forms that are in principle indefinitely recycleable. Nuclear power contributes directly to such an effort. As atomic energy furnishes more of the nation's heat and electricity, fossil stocks may be diverted from America's fuel inventory to the manufacturing base.

Might our descendants a thousand years hence prefer to deal with a continuing problem of radwaste maintenance, while enjoying the continued availability of conserved fossil feedstocks to its industrial base? Or might they prefer an environment free of long-lived reactor wastes—but also barren of essential petrochemical feedstocks, these literally having been turned into smoke by prior generations in order to avoid reliance on atomic energy? To such questions there can be no definitive answers. Yet some sensitivity to the issues raised thereby may still provide a useful perspective from which to consider the profounder problems of nuclear policy.

Electric Demand by Co-Located Industry

The gains to be achieved by co-locating firms with large electrical requirements may be less obvious—but they are not less real—than are those to be gained by co-locating heat-intensive industrial plants.

Electricity does not degrade with distance, as do steam and hot water. But line losses occur in electric transmission, contributing to cost increases as a function of the user's remoteness from the generator. As was pointed out in Chapter 5, the major economic disadvantage to the centering of electric capacity in power parks derives from the long-haul cost penalty. Every mile between the consumer of electricity and the source increases outlays for rights-of-way; increases the unsightliness of power lines; and increases vulnerability to line breakage, short circuiting, or sabotage.

Most of the electric load expected to exist in the year 2000 will not have materialized by 1980, when the first nuclear centers may be under construction. Some electricity from the early units in these power parks will indeed require transmission over long distances to already existing load centers in and near big cities. Consumers in these cities will have to bear the costs of long-haul delivery.

But with satellite and lattice siting, careful advanced planning can reduce (even if it cannot totally eliminate) the transmission penalty for firms that locate in the immediate vicinity of generating units foreseen to come on line in the future, as in Figure 8-2. Of course, in order to gain the cost benefits implied by Figure 8-2, both developers of the region's generating capacity and industrial firms whose decisionmakers anticipate plant expansion over future decades would have to know that the needed land for small industrial parks will be available years in advance. The needed assurances can be based only on a comprehensive regional plan.

Long-range planning offers significant opportunities to promote energy conservation. Inefficiency in electric generation not only raises power costs but also wastes energy. The older, less efficient plants in a utility's overall generating system are usually used to produce electricity during peak service hours. The newer, generally bigger and more efficient units serve the "base load." But an unanticipated jump in demand within a utility's service area generally necessitates increased use of the less-efficient peaking units to serve the expanded base load. A regional plan should permit industry to choose optimal sites within a

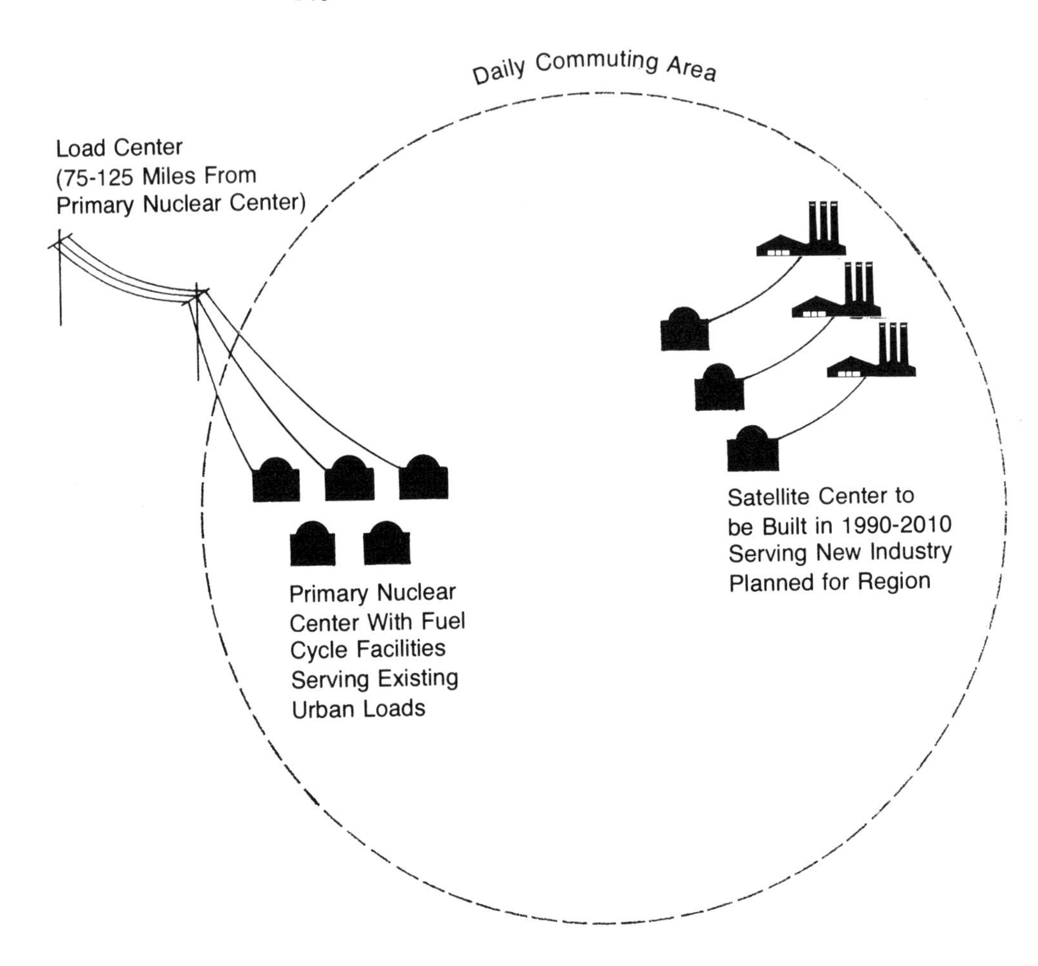

Figure 8-2. Satellite Siting with Advanced Planning to Industrialize.

Notes to Figure 8-2. The primary nuclear center in a region "targeted" for long-term economic growth would be constructed within a regional commuting system. Ideally, this primary center would be sited as close as possible to the boundary of the regional commuting system, and in the direction of the primary load center—within 125 miles or so of the major load that it serves. Beyond this limit, transmission costs can outweigh any on-site economic advantages of clustering reactors in a park. Yet the primary nuclear center would be located more than 60 to 75 miles from the center of its nearest main load center, and hence beyond ready commuting distance thereof.

If it were known on the basis of a comprehensive regional plan that a satellite nuclear center were later to be built at an indicated site some distance away from the primary nuclear center, industrial decisionmakers could choose to locate plants so as to minimize the costs of transmitting electricity. In this scheme, the cost of electricity from the primary nuclear center must reflect the transmission penalty that results from locating generators between 60 and 125 miles from existing load centers. But the cost of electricity from the eventual satellite center need not bear a major transmission penalty if industrial units capable of absorbing large quantities of bulk power locate in the region, effectively becoming a new load center taking power directly from the satellite.

broadly defined multicounty area. Reciprocally, such a plan would permit a regional G-T to develop its generating capacity on an optimal schedule to match a growing base load with appropriately energy-conservative base-loaded generating units.

The Optimal Size of a Regional City

Even in the absence of extensive industrialization, a power park would require the development of some regional infrastructure to serve construction workers, the nuclear plant operating force, and their families. *Infrastructure* refers to those built-in features of the environment which may be generally shared by all citizens—sewer lines, service roads, school buildings. Infrastructure can also refer to governmental decisionmaking bodies, police forces, sanitation units, and the like.

Under almost no circumstances could the number of construction workers peak below 4000, even for a small power park. General Electric estimates for a 20-unit nuclear center with an adjacent fuel cycle facility and some co-located industrial development suggest a workforce twice that size.[12] On the distant intermetropolitan frontier, additional workers would probably be needed to build new houses and the regional infrastructure.

A worker moving to the region would bring, on the average, two or three additional individuals (spouse and one or two children), implying an influx of up to 10,000 new families with a further need for additional services, from barber shops and beauty parlors to supermarkets and hospitals.[13] Thus a nuclear center could require the creation of a series of new communities (or the expansion of established communities) with a net population increase of some 50,000 people. When concentrated at a single location, this 50,000 population figure corresponds to the level at which a growing city begins to exert significant economic spread effects on the surrounding region—a fact that underscores the importance of "size of place" in predicting the socioeconomic characteristics of a given urban settlement.

A population influx of this size, combined with the promise of permanent employment, implies the possibility of building a true regional city, rather than relying on trailer towns to house a gypsy workforce. But can power park development support a regional city big enough to be viable both socially and economically?

Research since World War II suggests that the optimum size of an American city may be much smaller than had once been supposed. Much of the pertinent work has been summarized by Robert Dahl:

> There is . . . no worthwhile evidence that there are significant economies of scale in city government for cities over about 50,000 . . . . Per capita city expenditures increase with the size of the city, at least in the United States.[14]

Higher expenditures often evidence not higher levels of service but rather the *dis*economies of big cities. Thus rising costs may be disproportionately accounted for by a growing need to suppress crime or to service mass transit systems that have suffered from poor maintenance. The straitened condition of New York City in the late 1970s emerges as the apposite horrible example. Dahl continues:

> The oft-cited cultural advantages of metropolis are also largely illusory. On the basis of his research on American cities, Duncan estimates that the requisite population base for a library of "desirable minimum professional standards" is 50,000-75,000, for an art museum, 100,000. . . .[15]

The optimum city size, Dahl concludes, falls in the population range of 50,000 to 200,000.

Independent studies by researchers at the Urban Institute in Washington, D.C. confirm Dahl's contention that both capital and operating costs for public services are minimized (without losses in quality of service) for settlements in the range of 50,000 to 200,000. Costs are higher for cities above this level, as well as for villages below 50,000. Moreover, according to Thomas Muller of the Urban Institute, even the *rate* of cost escalation during the inflationary early seventies seemed to be under most satisfactory control in small cities.[16] Of course, these analyses abstract from the problems of big cities that trace to fragmented or corrupt political authorities. More important, the Dahl and Muller surveys relate primarily to efficient expenditures in the public sector. A more complete treatment of the "optimum size" problem must reckon with the fact that private commercial or manufacturing activities often benefit from the scale economies and reduced transfer costs implicit in cities much larger than communities on America's intermetropolitan frontier would be likely to become.

Power park development probably can support the emergence of settlements within the range of optimal public effectiveness and, on the evidence of survey data reported in Chapter 6, optimal social desirability. This population level—from 50,000 to 200,000—may prove, however, to be suboptimal for long-term economic growth driven by entrepreneurial activity in the private sector. So the question occurs: In the event that the citizens in a regional city decide on further growth, can they reasonably expect to adopt policies favoring a full explosion into cityhood?

The Explosion into Cityhood

No consensus has emerged among economists to suggest any single or sure path to regional growth. Broadly speaking, however, students of economic development divide into two categories. The first includes those who focus on "export base" theories; the second, those who emphasize "import substitution."

Export Base versus Import Substitution

Export base theorists generally reason that the ability of merchants in a given region to sell goods beyond their own borders—whether these goods be raw materials such as oil or a manufactured product such as electricity—will gain a favorable balance of trade for the region in question. Exports bring a surplus of money, called "exchange." This exchange may then be used to purchase needed goods from external supply areas or else to finance the further internal development of the region. In either event, prosperity depends on an ability successfully to compete in the export market.

Export base theorists emphasize the ratio between employment in the region's export industry (i.e., the power park, which exports electricity for sale beyond the multicounty impact area) and total regional employment. If every worker in an export-related job supports a specific fraction of the business of the region's supporting workers—grocers, bankers, doctors, and so forth—then the ratio of export-related employment to total regional employment will be fixed. This ratio will invariably exceed unity. So an increase in jobs created by an expansion of the region's export base will produce a rise in demand for supporting services.

A power park on the intermetropolitan frontier would itself be a hugely capitalized, one-product export base built to send huge quantities of electricity to distant load centers. The potential of such a facility to serve as a regional growth pole will depend on the uses to which the cash flow during construction, as well as the subsequent profits from bulk electric sales, are put. Power parks will stimulate regional development only if a fair share of the revenues ultimately to be derived from the export of electricity are intelligently invested *within the region.* Hence it must lie with the authors of the regional plan to ensure that the park will not become a kind of colonial outlier of distant financial powers. The locale should not bear the socioeconomic strains and environmental risks only to see the long-term fiscal returns flow to moneymen outside the region. Ways must be found to anticipate future revenues, to "borrow against them," and to use these monies for the development of the region's own infrastructure. Infrastructure is essential for further industrialization. The ability to attract additional industry to the region in turn emerges as the key to supplementing an export base strategy with policies that emphasize import substitution.

Import substitution theorists stress the importance of local manufacturing firms that decrease the region's reliance on external sources of finished goods.

We saw in Chapter 5 that the costs of long-haul transmission may undercut much of the economic case for power parks. An excessive cost penalty would in some areas, especially in the West, argue against siting far out on the intermetropolitan frontier. The incentive to cut costs by shortening transmission lines, however, needs to be weighed against the desirability of siting parks away from builtup areas in order to provide growth poles for receding areas of rural America. Suppose that sites where the penalty of long-haul transmission costs just balances the expected gains of a growth pole strategy exist at significant distances from major urban centers. Then the transportation costs of importing goods into the region of the power park will provide an incentive to develop indigenous regional industries serving the multicounty market defined by the regional commuting system of the power park.

The development of import substitution policies on the intermetropolitan frontier can occur only gradually, since throughout the early phases of regional development, communities in the neighborhood of a nuclear center would have to rely on external suppliers. In order to sustain both a growing population and the industries that employ them, finished goods must be imported in wide variety from areas outside the locale: bicycles from Toledo (or Germany), watches from New Haven (or Switzerland), radios from Chicago (or Japan), and so forth.

A new stage of development begins when entrepreneurs in the region recognize opportunities for profitable local manufacture of needed goods. Raw materials may perforce continue to be imported. (In the event of nuclear center siting near America's fossil reserves to gain feedstocks for petrochemical and refining processes, the main raw materials already lie within the region.) These raw materials will be fashioned into a growing array of locally produced finished goods. Such goods may then be substituted for manufactured products that previously had to be imported. In time, a resident may be able to buy a locally produced bike or a watch made in a factory now down the street.

When substantial import substitution begins to occur, the natural history of a developing community passes into the phase of explosive economic development—the explosion into cityhood that brings the developmental process full term.

Just as a series of towns may result from an extended process whereby industries agglomerate to a nuclear center in order to internalize certain externalities (e.g., by capturing waste heat or taking advantage of a skilled ready-made workforce), the explosion into cityhood occurs when entrepreneurs begin to internalize the town's once external supply hinterland. To do this, local leaders create a new supply base that becomes part of the economic structure of the immediate locale.

Thus the logic of agglomeration runs its course. First construction of a power park creates local conditions attractive to firms that need heat, electricity, or skilled workers as inputs to their own processes. The workers at the construction site and at these agglomerating firms, plus their families and service personnel, constitute a growing market for consumer goods. Responding to the needs of this local market, additional firms set up business in the region–firms such as that of the watch factory or cycle manufacturer, which do not show any special dependency on the power park except insofar as the park helps create a growing regional market.

Import substitution may be regarded as the obverse side of the development of a favorable export base. Profits from the export trade may be capitalized within the immediate locale of the export base instead of being used to buy imports. That is, instead of being "expensed off" to purchase goods from outside the region, exchange may be converted into local investments. The manufacturer of goods within the locale becomes a neighbor rather than some faceless entrepreneur from a distant city. Thus with import substitution, more and more of the wealth ultimately generated by the export of electricity remains within the area of the spread effects of the power park. The retained regional profits, together with the transportation costs that are saved by "buying local," become effective increments to the economic base of the local community. Judicious investments of these increments can help maintain the developmental spiral. With the explosion of economic activity triggered by import substitution, the dynamics of commercial displacement within the region become more pronounced.

"Tuning" Regional Growth

An approach to regional development that combines attention both to the export base and to import substitution highlights transportation as critically important. The explosion into cityhood may be guided by controlling the development of the transportation network that will serve the communities in the vicinity of the power park. Especially necessary with satellite siting and lattice siting is a regional roadway system so workers may commute from dispersed residential settlements. Such regional commuting systems might follow the model of those which have emerged in metropolitan areas. C.A. Doxiadis' and Brian J.L. Berry's concept of the "daily urban system" refers to the area within which substantial numbers of people may regularly travel to and from work in a single day.[17] In the Northeast, for example, interstate parkways and commuter rail lines make daily round trips in excess of 100 miles commonplace.

It bears mention that considerable research remains to be done before the full dimensions of the "regional city" concept can be appreciated, or all the possible problems of this concept understood. A regional city would lack the diversity of a typical metropolis and hence could function as a coherent social entity only if internal transportation facilities permit efficient transfers of people and goods within the region. New road construction, however, is a notoriously costly kind of infrastructure development. A million dollars per mile of surfaced highway is a commonly used figure. Hence economic constraints limit the perimeter of a region to be served by newly laid roads. The radius of this perimeter in turn would limit the dispersion of towns within commuting distance of job sites in the region.

The regional *commuting* system must be distinguished from the interregional *arterial* network. Although a well-conceived commuting system contributes to the internal integration of a region, it need not contribute significantly to the linkup of the region to the national economy. Such a linkup depends on the transportation arteries that connect cities and industries to markets or supply centers beyond the contours of the daily commuting system.

Emphasizing the regional commuting system at the expense of the arterial transportation network can stimulate internal development through import substitution. Favoring arterial over regional commuting transport works in the other direction. For as the arterial network grows in efficiency and complexity, transportation costs diminish as a factor in the prices of imported goods. An initially remote locale on the intermetropolitan frontier will then become more effectively integrated into the national market. Thus, development of arterial transportation routes tends to dampen the incentives for continuing import substitution, thereby also dampening the dynamics of continuing explosive population growth and economic activity.

Coda: "Hope Enough and To Spare"

The emergence of a city of the second sun should follow an anticipated sequence, which it would be the purpose of the regional plan to guide and perhaps in some particulars to modify. The pace of regional population growth (first as a result immigration by construction workers and their families, then from rising natality within the region) determines the speed with which the emerging regional city will move through the successive phases of its natural development. Economic incentives, local land-use policies, and zoning statutes—all consistent with the terms of the regional plan—can be used to "tune" the growth process. Thus may public officials redirect, accelerate, retard, or even arrest the natural history of a developing community.

The long-term results may be more than merely the creation of power parks, or even a complex of new towns with elaborated transportation systems in America's rural hinterland. A power park with even moderate associated industrial development could easily add to a sparsely populated region an increment within the 50,000 to 200,000 population range. But instead of concentrating in one urban center, a population increment of this size would be dispersed among a series of towns, forming a single regional city within the contours of a unified commuting system. Such regional cities would develop over several decades according to regional plans—becoming in the early twenty-first century, when fusion reactors start coming on line, true cities of the second sun in areas of the land that badly need them.

The optimistic suggestion that two dozen or more regional cities could be developed "according to plan" does *not* mean that nuclear fission is without problems, or that the promise of fusion power is assured. It does *not* imply that power parks are superior in all respects to dispersed siting. It does *not* even suggest that an incontrovertible case can be made for a growth pole approach to nuclear center planning.

But the foregoing chapters do suggest that, on balance, the gains of a growth pole strategy based on nuclear centers would probably exceed the costs. Admittedly, to gamble on cities of the second sun could put the health of whole populations and the environmental quality of whole regions at some risk. But to pass up the gamble would carry greater risks–the risk of impoverishing the nation's energy base, the risk of subjecting future generations to simultaneous fossil pollution and fossil depletion, and the risk of foregoing opportunities to rescue needy regions of the land from secular economic decline.

Clustering nuclear plants as growth poles in rural America would not only be risky but also revolutionary–in the original sense of that word. Such a policy would represent a return to earlier values and forms, a reversion to the original American tradition of expansion ever beyond the margins of existing population concentrations. As symbols of America's frontier ethic, cities of the second sun will require more than physical developments (i.e., the building of nuclear complexes) and more even than carefully designed supporting social programs (i.e., the raising of new towns or the planned growth of nearby existing communities).

Cities of the second sun imply an attitude of confident experimentation. America began with such a mood. Even as he championed an English program of overseas plantations during the time of Elizabeth I, Francis Bacon admitted that formidable obstacles stood in the way of successful colonization of the New World. But, he argued,

> There is no comparison between that which we may lose by not trying and by not succeeding. . . . There is hope enough and to spare, not only to make bold men try, but also to make a sober-minded and wise man believe.[18]

Hope enough and to spare should be mustered by those who have an opportunity simultaneously to meet the energy needs of twenty-first century America and to recover the tradition of town building on America's frontier.

Notes

1. *The Economy of Cities* (New York: Random House, 1969), Chap. 5.
2. Marvin Yarosh, ed., "Waste Heat Utilization: Proceedings of the

National Conference" (October 27-29, 1971, Gatlinburg, Tennessee), available as CONF-711031 from the National Technical Information Service, U.S. Department of Commerce. Particularly useful are the papers by Merle Jenson, William Yee, A.J. Miller, and William Kessler, all in the "Technical Overview" of Section I; as well as the reports on specific experimental programs by George Vanderborgh, Gerald Williams, and W.R. Watts, all in "Demonstration Projects," Section III. See also Ronald Stewart and S.P. Mathur, "Handling Hot Water With a Payoff," *Conservationist* (December-January 1970-71), 16-20.

3. On the technology of the heat pump, see Wilson Clark, *Energy for Survival* (Garden City: Doubleday, 1972), pp. 245ff.

4. Lars Elemenius, "District Heating in Sweden," reported in Anton Schmalz, ed., *Energy: Today's Choices, Tomorrow's Opportunities*, Proceedings of the World Futurist Society Conference, Hilton Hotel, Washington, D.C., February 19-21, 1975.

5. "Conceptual Design of a Fusion Power Plant to Meet the Total Energy Requirements of an Urban Complex," presented at the Nuclear Fusion Reactor Conference at Culham Laboratory, England (September 1969), p. 9.

6. See generally, G.E. Assessment, Vol. II, pp. 5-1ff.

7. Report by the N.G.C. Center for Policy Research and Analysis, prepared in conjunction with the Southern Interstate Nuclear Board, "Power Parks From the States' Perspective" (March 1975), p. 3.

8. Similarly in Sam Schurr and Jacob Marschak, *Economic Aspects of Atomic Power* (Princeton: Princeton University Press, 1950), one of the first assessments of the potential of nuclear power as a tool of industrial development. See especially Part Two and Chap. 13.

9. See the analysis of the LWR-MSF system in "Nuclear Energy Centers . . . ," ORNL-4290, pp. 34-40. J.W. Michel summarized subsequent work at Oak Ridge National Laboratory on the agroindustrial complex concept in "Status and Recent Developments in Agro-Industrial Complex Studies" (ORNL-TM-319, June 1, 1970). See especially at pp. 23-24. See also William E. Hoehn, "Prospects for Desalted Water Costs" (Santa Monica: The Rand Corporation, RM-5971-FF, October 1969); and Frederick J. Wells, "Review and Analysis of Costs of Desalted Sea Water" (Office of Water for Peace, U.S. Department of State, 1969).

10. See Morris Beller et al., "Applications of Fusion Power Technology to the Chemical Industry" (Brookhaven National Laboratory, BNL-18815, June 2, 1974), especially at pp. 12ff.; and The Dow Chemical study, "Industrial Complex Study for the Wasatch Front Area, State of Utah" under Contract 341, Utah Department of Natural Resources (Midland, Michigan, May 17, 1973), *passim*, and especially at pp. 43ff.

11. Meyer Steinberg and Seymour Baron, "Synthetic Carbonaceous Fuel and Feedstock Using Nuclear Power, Air and Water" (January 1976), draft of a

paper for the First World Hydrogen Energy Conference, March 1-3, 1976, University of Miami Clean Energy Institute.

12. See the estimates and manpower programs summarized in my report for the G.E. Assessment, Vol. II, at p. 6-133; see also NECSS, Part I, pp. 4-54ff.

13. Human Affairs Research Center, Battelle Institute, "Identification and Management of Economic and Social Impacts of Nuclear Energy Centers: A Preliminary Analysis" (Battelle Institute, Pacific Northwest Division, September 1975), pp. 19-26. See also in general on socioeconomic impacts, Department of Housing and Urban Development's *Rapid Growth from Energy Projects* (HUD-CPD-140, April 1976), especially "Employment and Population" at pp. 3-12; and the longer overview pamphlet by Centaur Management Consultants, *Managing the Social and Economic Impacts of Energy Developments* (Washington, U.S. ERDA, July 1976), *passim.*

14. Presidential Address to the American Political Science Association "The City in the Future of Democracy," 61 *American Political Science Review* (December 1967), 953, p. 965.

15. Ibid., p. 966.

16. "Growing and Declining Urban Areas: A Fiscal Comparison," (Washington, D.C. The Urban Institute, November 1975), especially at p. 23.

17. See generally the useful and suggestive papers by Jerome A. Pickard (pp. 127-184), and Brian J.L. Berry (pp. 229-265), especially at pp. 233ff., in Sarah Mills Mazie, ed., *Population, Distribution and Policy*, Volume V of the Report of the Commission on Population Growth and the American Future (Washington, U.S.G.P.O., 1972).

See also John Friedman and John Miller, "The Urban Field," 31 *Journal of the American Institute of Planners* (November 1965) 312; and Brian J.L. Berry, *Metropolitan Area Definition* (Census Bureau Working Paper No. 28, U.S.G.P.O., 1968).

18. Quoted from Bacon's *New Atlantis* in Catherine Drinker Bowen, *Francis Bacon: Temper of a Man* (Boston: Houghton Mifflin, 1962), p. 2.

Additional References and Acknowledgments

The best general introduction to nuclear technology known to me will be found in the section entitled "Nuclear Fission as an Energy Source" in Lon C. Ruedisili and Morris W. Firebaugh, *Perspectives on Energy* (Oxford University Press, 1975). Particularly useful are: Manson Benedict, "Electric Power from Nuclear Fission," p. 196; David Rose, "Nuclear Electric Power," p. 209; and A.S. Kubo and David Rose, "Disposal of Nuclear Wastes," p. 238. See also the September and October 1971 issues of the *Bulletin of Atomic Scientists*, devoted entirely to articles on "The Energy Crisis," and the July-August *Skeptic* symposium volume, "Nuclear Energy: Do the Benefits Outweigh the Risks?"—particularly the articles by David Dinsmore Comey and Richard Kipling.

An exceptionally fine survey of quantitative aspects of the nuclear scene—albeit one biased in favor of atomic power—is Vance L. Sailor and Ferdinand Shore's *The Future of Nuclear Power in the Northeast* (Brookhaven National Laboratory BNL-50551, March 20, 1976), especially Chapters III, IV, and VII. Other surveys, useful as background to the materials in Chapters 1, 2, and 4 of the foregoing analysis, are Dixy Lee Ray, *The Nation's Energy Future* (U.S. Atomic Energy Commission, December 1, 1973, WASH-1281), pp. 107-113; and the U.S. Energy Research and Development Administration's *National Plan for Energy Research, Development and Demonstration* (June 30, 1976, ERDA-76-1), Vol. 2, pp. 257-311 and, in connection with thermonuclear fusion power, pp. 199-218.

For readable and reasonably balanced accounts of the antinuclear position, see McKinley Olson, *Unacceptable Risk* (Bantam, 1976), especially Chapters I-II; and Walter Patterson, *Nuclear Power* (Baltimore: Penguin, 1976), Chapters 7-8. See also the National Council of the Churches of Christ, "The Plutonium Economy" (October 10, 1975), available from the National Council; Hannes Alfven, "Fission Energy and Other Sources of Energy," *Science and Public Affairs* (January 1974); and two articles by James G. Phillips: "Nader, Nuclear Industry Prepare to Battle Over the Atom," 7 *National Journal Reports* (February 1, 1975), pp. 153-164; and "Safeguards, Recycling Broaden Nuclear Power Debate," 7 *National Journal Reports*, (March 22, 1975), pp. 419-429.

A useful comparative analysis of the economic aspects of alternative electric generating modes appears in R.K. Pachauri, *The Dynamics of Electrical Energy Supply and Demand* (New York: Praeger, 1975), especially at pp. 84-113.

Generally useful in connection with nuclear cost problems are Edward J. Mitchell, ed., *Energy Self-Sufficiency: An Economic Evaluation by the M.I.T. Energy Laboratory Policy Study Group* (Washington, D.C., November 1974), pp. 45-48; and B.E. Prince et al., "Nuclear Fuel Cycle Economics: 1970-1985," for the Office of Energy Systems Analysis, U.S. ERDA, October 24, 1975 (Oak Ridge National Laboratory, under contract W-7405-eng-26).

On nuclear health effects, Herman Somers, "Atomic Energy and Workmen's Compensation," before the Pacific Coast Metal Trades Council Conference, May 1957, Long Beach, remains a lucid statement of general principles. Statistical data appear in: Federal Radiation Council Staff Report No. 8, *Guidance for the Control of Radiation Hazards in Uranium Mining* (September 1967); and N. Spencer et al., "Control of Radiation Exposure in Uranium Mines; A Cost and Economic Analysis" (November 1968)–both available through the U.S. Federal Radiation Council; and *Radiation Standards for Uranium Mining*, Hearings before the Subcommittee on Research, Development and Radiation of the Joint Committee (March 17-18, 1969), U.S.G.P.O.

On reactor safety, see Frank von Hippel's "Nuclear Reactor Safety," 32 *Bulletin of the Atomic Scientists* (January 1976), 44. Also useful have been: Walter Jordan, "Nuclear Energy: Benefits versus Risks," *Physics Today* (May 1970), 32; and Chauncey Starr, "Social Benefit versus Technological Risk," 165 *Science* (September 19, 1969), 1232. These articles are rich in quantitative data. Less quantitative data are John T. Edsall, "Hazards of Nuclear Fusion Power and the Choice of Alternatives," 1 *Environmental Conservation* (Spring 1974), 21; and "Inside an Atomic Power Plant–How Dangerous?" *U.S. News and World Report* (March 15, 1976), 61.

A concise yet comprehensive survey of the organizational and commercial environment of the evolution of the American nuclear industry will be found in John F. Hoggerton's "The Arrival of Nuclear Power," in Gene I. Rocklin, ed., *Scientific Technology and Social Change* (San Francisco: W.H. Freeman, 1974), 283. The authoritative study of the manpower aspects of the nuclear industry–particularly illuminating for its discussion of personnel procedures within the electric utilities and in the major equipment companies such as General Electric and Westinghouse–is James W. Kuhn's *Scientific and Managerial Manpower in Nuclear Industry* (New York: Columbia University Press, 1966). Chapters II through IV of the Kuhn study have proved particularly useful.

In 1974, a "Federal Energy Regulation Study Team" under the Chairmanship of William D. Doub, an AEC Commissioner, completed a study and evaluation of the U.S. regulatory process, focused on the federal level. The report, *Federal Energy Regulation: an Organizational Study*, was released in April, 1974, and is available through the U.S. Government Printing Office. I have drawn on this source and also extensively on materials gleaned from three unpublished papers, all prepared by students of the regulatory process in the course of research on the power park concept: Abraham Braitman, "Antitrust Aspects of Electric Facility Siting" (Office of Antitrust and Indemnity, U.S. Nuclear Regulatory Commission, no date); Neal Carothers, III, "Organization for Electric Energy Supply" (National Academy of Public Administration, March 1975); and Arvin Upton and E. David Doane, "Current Regulatory Requirements Which Would Be Applicable to Energy Parks" (LeBoef, Lamb, Leiby and MacRae, Washington, D.C., February 18, 1975).

I have also found the following two discussions of the National Environmental Protection Act, both by Judith A. Best of the Cornell University Energy Project, very helpful; "NEPA: Some Legal Constraints as Set Forth by the Court in *Calvert Cliffs*" (October 1971), and "New Institutional Arrangements to Resolve Power Plant Siting Conflicts: A Political Analysis" (February 1972). These papers are available from the Cornell Energy Project, Ithaca, New York.

In connection with the discussions in Chapters 6 and 8, the following standard references on regional economic and regional development theories may usefully be consulted: John Cumberland, *Regional Development Experiences and Prospects in the United States of America* (The Hague: Mouton, 1973); Benjamin Higgins, *Economic Development* (New York: W.W. Norton, 1968); Walter Isard, *Methods of Regional Analysis* (Cambridge, Mass.: M.I.T. Press, 1960); David McKee, Robert Dean, and William Leahy, eds., *Regional Economics* (New York: The Free Press, 1970); Hugh Nourse, *Regional Economics* (New York: McGraw-Hill, 1968); Harry Richardson, *Regional Economics* (New York: Praeger, 1969).

Perhaps the best known center of regional research is Walter Isard's group at the University of Pennsylvania. Members of this group prepared a comprehensive study of the probable regional development impacts of power parks under the title, "Regional Economic Impacts of Nuclear Energy Centers" for the U.S. Nuclear Regulatory Commission (mimeo, September 30, 1975).

The following document suggests ways in which a "growth pole" strategy might dovetail with programs of the Economic Development Administration: U.S. Department of Commerce: Program Evaluation: *The Economic Development Administration Growth Center Strategy* (Washington, D.C., February 1972). See, for a further discussion and evaluation of the growth pole approach to regional development, Edgar M. Hoover, *An Introduction to Regional Economics* (New York: Alfred A. Knopf 1971, 277; Niles Hansen, "Development Pole Theory in a Regional Context," in McKee, Dean, and Leahy, eds., *Regional Economics*, 128; and also Hansen's *Location Preferences, Migration, and Regional Growth* (New York: Praeger, 1973), especially at pp. 15-20.

The following may be taken as representative of a large (and growing) body of literature that challenges the thesis that economic growth is necessarily either desirable or dependent on major investments in central, capital-intensive industrial projects: E.J. Mishan, *The Costs of Economic Growth* (New York: Praeger, 1967), Dennis Meadows et al., *The Limits to Growth* (Cambridge, Mass.: M.I.T. Press, 1972); Herman E. Daly, ed., *Essays Toward a Steady-State Economy* (San Francisco: W.H. Freeman, 1972).

The study of "socioeconomic impacts" has become a favorite indoor sport of recent years among social scientists and government officials. See (in addition to the HUD and ERDA handbooks referenced in the notes to Chapter 8) *Urban and Rural America: Policies for Future Growth*, report by the Advisory Commission on Intergovernmental Relations (Washington, D.C.: U.S.G.P.O.,

April 1968), especially pp. 73-77, 103-105, Chapter VI; Berkshire County Regional Planning Commission, Karl Hekler, director, *Evaluation of Power Facilities: A Reviewer's Handbook* (Pittsfield, Mass., April 1974 under contract to HUD), especially Chapter 8; Applied Physics Laboratory, Johns Hopkins University, Policy Research Associates, *A Regional Pilot Study for the National Energy Siting and Facility Report* (Available as APL/SHU, CP 043, February 1976), especially Chapter VII. Readers desiring a comprehensive bibliography should consult the references given in the Human Affairs Research Center of Battelle Northwest, "Identification and Management . . . " listed in the notes to Chapter 8.

Useful data are also to be found in the so-called confirmatory assessment studies of socioeconomic impacts of nuclear power plant construction projects by the Cost-Benefit Analysis Branch, Office of Nuclear Reactor Regulation, U.S. Nuclear Regulatory Commission. See the studies by Lynn Pollucow (McGuire Nuclear Plant, N.C.), October 22, 1975; and by Suzanne Pollucow (McGuire Nuclear Plant, Ark.), December 17, 1975. See also "Resource City, Rocky Mountains," a study of socioeconomic impacts prepared by the Federation of Rocky Mountain States, Inc., Denver, Colorado (in mimeo, no date); and the study by C.R. Cavanagh of Stone and Webster Engineering Corp., "Community Development Implications of Energy Complex Construction" (no date), furnished by the author, and particularly helpful in connection with the analysis of socioeconomic impacts.

Also pertinent to the issue of the social impacts of a power park is the study, "The Impact of a Proposed Synthetic Crude Oil Project on Fort MacMurray," by Reed, Crowther & Partners, Consulting Engineers, for Syncrude Canada (February 1973). This study provides quantitative estimates of the probable consequences of a development equivalent to a small park or a small city, broken out to show such categories of effect as recreational housing requirements, and the like. Similarly useful is "Construction City Study for a Nuclear Power Center," prepared by United Engineers and Constructors, J.H. Prowley, Project Manager, for the U.S. Atomic Energy Commission (January 15, 1974). I have used this study, which includes a long illustrated appendix on experiences to date with planned communities in the United States and abroad, as a basic reference document for the analysis of the "new town" aspects of energy parks. "Financing New Communities–Government and Private Experience in Europe and the United States" (1973), available from the Office of International Affairs, U.S. Department of Housing and Urban Development, describes the wide variety of fiscal techniques that are available to developers of planned communities. Accounts of specific case studies–e.g., Columbia, Maryland, the British New Towns programs–are included.

Also recommended is V.G. Davidovich, *Town Planning in Industrial Districts* (Copyright 1968, Israel Translation Program), a treatise on city development in the Soviet Union, available from the U.S. Department of Commerce.

Index

Index

About the Author

Gerald Garvey is professor of politics, Center of International Studies, Princeton University, where he serves as an editor of *World Politics* and as comaster (with his wife, Lou Ann Benshoof Garvey) of Princeton Inn College. Professor Garvey has served in a variety of governmental posts, including as director of planning, Federal Power Commission and, in 1973, as chairman, New Jersey Energy Crisis Commission. He has published widely in scholarly journals, including the *American Political Science Review*, the *Harvard Journal on Legislation*, the *Public Administration Review*, the *Technology Review*, and *World Politics*. His books include *International Resource Flows* (edited, with Lou Ann Benshoof Garvey); *Energy, Ecology, Economy*; and *Constitutional Bricolage*.

Center of International Studies: List of Publications

Gabriel A. Almond, *The Appeals of Communism* (Princeton University Press 1954)

William W. Kaufmann, ed., *Military Policy and National Security* (Princeton University Press 1956)

Klaus Knorr, *The War Potential of Nations* (Princeton University Press 1956)

Lucian W. Pye, *Guerrilla Communism in Malaya* (Princeton University Press 1956)

Charles De Visscher, *Theory and Reality in Public International Law*, trans. by P.E. Corbett (Princeton University Press 1957; rev. ed. 1968)

Bernard C. Cohen, *The Political Process and Foreign Policy: The Making of the Japanese Peace Settlement* (Princeton University Press 1957)

Myron Weiner, *Party Politics in India: The Development of a Multi-Party System* (Princeton University Press 1957)

Percy E. Corbett, *Law in Diplomacy* (Princeton University Press 1959)

Rolf Sannwald and Jacques Stohler, *Economic Integration: Theoretical Assumptions and Consequences of European Unification*, trans. by Herman Karreman (Princeton University Press 1959)

Klaus Knorr, ed., *NATO and American Security* (Princeton University Press 1959)

Gabriel A. Almond and James S. Coleman, eds., *The Politics of the Developing Areas* (Princeton University Press 1960)

Herman Kahn, *On Thermonuclear War* (Princeton University Press 1960)

Sidney Verba, *Small Groups and Political Behavior: A Study of Leadership* (Princeton University Press 1961)

Robert J.C. Butow, *Tojo and the Coming of the War* (Princeton University Press 1961)

Glenn H. Snyder, *Deterrence and Defense: Toward a Theory of National Security* (Princeton University Press 1961)

Klaus Knorr and Sidney Verba, eds., *The International System: Theoretical Essays* (Princeton University Press 1961)

Peter Paret and John W. Shy, *Guerrillas in the 1960's* (Praeger 1962)

George Modelski, *A Theory of Foreign Policy* (Praeger 1962)

Klaus Knorr and Thornton Read, eds., *Limited Strategic War* (Praeger 1963)

Frederick S. Dunn, *Peace-Making and the Settlement with Japan* (Princeton University Press 1963)

Arthur L. Burns and Nina Heathcote, *Peace-Keeping by United Nations Forces* (Praeger 1963)

Richard A. Falk, *Law, Morality, and War in the Contemporary World* (Praeger 1963)

James N. Rosenau, *National Leadership and Foreign Policy: A Case Study in the Mobilization of Public Support* (Princeton University Press 1963)

Gabriel A. Almond and Sidney Verba, *The Civic Culture: Political Attitudes and Democracy in Five Nations* (Princeton University Press 1963)

Bernard C. Cohen, *The Press and Foreign Policy* (Princeton University Press 1963)

Richard L. Sklar, *Nigerian Political Parties: Power in an Emergent African Nation* (Princeton University Press 1963)

Peter Paret, *French Revolutionary Warfare from Indochina to Algeria: The Analysis of a Political and Military Doctrine* (Praeger 1964)

Harry Eckstein, ed., *Internal War: Problems and Approaches* (Free Press 1964)

Cyril E. Black and Thomas P. Thornton, eds., *Communism and Revolution: The Strategic Uses of Political Violence* (Princeton University Press 1964)

Miriam Camps, *Britain and the European Community 1955-1963* (Princeton University Press 1964)

Thomas P. Thornton, ed., *The Third World in Soviet Perspective: Studies by Soviet Writers on the Developing Areas* (Princeton University Press 1964)

James N. Rosenau, ed., *International Aspects of Civil Strife* (Princeton University Press 1964)

Sidney I. Ploss, *Conflict and Decision-Making in Soviet Russia: A Case Study of Agricultural Policy, 1953-1963* (Princeton University Press 1965)

Richard A. Falk and Richard J. Barnet, eds., *Security in Disarmament* (Princeton University Press 1965)

Karl von Vorys, *Political Development in Pakistan* (Princeton University Press 1965)

Harold and Margaret Sprout, *The Ecological Perspective on Human Affairs, With Special Reference to International Politics* (Princeton University Press 1965)

Klaus Knorr, *On the Uses of Military Power in the Nuclear Age* (Princeton University Press 1966)

Harry Eckstein, *Division and Cohesion in Democracy: A Study of Norway* (Princeton University Press 1966)

Cyril E. Black, *The Dynamics of Modernization: A Study in Comparative History* (Harper and Row 1966)

Peter Kunstadter, ed., *Southeast Asian Tribes, Minorities, and Nations* (Princeton University Press 1967)

E. Victor Wolfenstein, *The Revolutionary Personality: Lenin, Trotsky, Gandhi* (Princeton University Press 1967)

Leon Gordenker, *The UN Secretary-General and the Maintenance of Peace* (Columbia University Press 1967)

Oran R. Young, *The Intermediaries: Third Parties in International Crises* (Princeton University Press 1967)

James N. Rosenau, ed., *Domestic Sources of Foreign Policy* (Free Press 1967)

Richard F. Hamilton, *Affluence and the French Worker in the Fourth Republic* (Princeton University Press 1967)

Linda B. Miller, *World Order and Local Disorder: The United Nations and Internal Conflicts* (Princeton University Press 1967)

Henry Bienen, *Tanzania: Party Transformation and Economic Development* (Princeton University Press 1967)

Wolfram F. Hanrieder, *West German Foreign Policy, 1949-1963: International Pressures and Domestic Response* (Stanford University Press 1967)

Richard H. Ullman, *Britain and the Russian Civil War: November 1918-February 1920* (Princeton University Press 1968)

Robert Gilpin, *France in the Age of the Scientific State* (Princeton University Press 1968)

William B. Bader, *The United States and the Spread of Nuclear Weapons* (Pegasus 1968)

Richard A. Falk, *Legal Order in a Violent World* (Princeton University Press 1968)

Cyril E. Black, Richard A. Falk, Klaus Knorr and Oran R. Young, *Neutralization and World Politics* (Princeton University Press 1968)

Oran R. Young, *The Politics of Force: Bargaining During International Crises* (Princeton University Press 1969)

Klaus Knorr and James N. Rosenau, eds., *Contending Approaches to International Politics* (Princeton University Press 1969)

James N. Rosenau, ed., *Linkage Politics: Essays on the Convergence of National and International Systems* (Free Press 1969)

John T. McAlister, Jr., *Viet Nam: The Origins of Revolution* (Knopf 1969)

Jean Edward Smith, *Germany Beyond the Wall: People, Politics and Prosperity* (Little, Brown 1969)

James Barros, *Betrayal from Within: Joseph Avenol, Secretary-General of the League of Nations, 1933-1940* (Yale University Press 1969)

Charles Hermann, *Crises in Foreign Policy: A Simulation Analysis* (Bobbs-Merrill 1969)

Robert C. Tucker, *The Marxian Revolutionary Idea: Essays on Marxist Thought and Its Impact on Radical Movements* (W.W. Norton 1969)

Harvey Waterman, *Political Change in Contemporary France: The Politics of an Industrial Democracy* (Charles E. Merrill 1969)

Cyril E. Black and Richard A. Falk, eds., *The Future of the International Legal Order*. Vol. I: *Trends and Patterns* (Princeton University Press 1969)

Ted Robert Gurr, *Why Men Rebel* (Princeton University Press 1969)

C. Sylvester Whitaker, *The Politics of Tradition: Continuity and Change in Northern Nigeria 1946-1966* (Princeton University Press 1970)

Richard A. Falk, *The Status of Law in International Society* (Princeton University Press 1970)

Klaus Knorr, *Military Power and Potential* (D.C. Heath 1970)

Cyril E. Black and Richard A. Falk, eds., *The Future of the International Legal Order*. Vol. II: *Wealth and Resources* (Princeton University Press 1970)

Leon Gordenker, ed., *The United Nations in International Politics* (Princeton University Press 1971)

Cyril E. Black and Richard A. Falk, eds., *The Future of the International Legal Order.* Vol. III: *Conflict Management* (Princeton University Press 1971)

Francine R. Frankel, *India's Green Revolution: Political Costs of Economic Growth* (Princeton University Press 1971)

Harold and Margaret Sprout, *Toward a Politics of the Planet Earth* (Van Nostrand Reinhold 1971)

Cyril E. Black and Richard A. Falk, eds., *The Future of the International Legal Order.* Vol. IV: *The Structure of the International Environment* (Princeton University Press 1972)

Gerald Garvey, *Energy, Ecology, Economy* (W.W. Norton 1972)

Richard Ullman, *The Anglo-Soviet Accord* (Princeton University Press 1973)

Klaus Knorr, *Power and Wealth: The Political Economy of International Power* (Basic Books 1973)

Anton Bebler, *Military Rule in Africa: Dahomey, Ghana, Sierra Leone, and Mali* (Praeger Publishers 1973)

Robert C. Tucker, *Stalin as Revolutionary 1879-1929: A Study in History and Personality* (W.W. Norton 1973)

Edward L. Morse, *Foreign Policy and Interdependence in Gaullist France* (Princeton University Press 1973)

Henry Bienen, *Kenya: The Politics of Participation and Control* (Princeton University Press 1974)

Gregory J. Massell, *The Surrogate Proletariat: Moslem Women and Revolutionary Strategies in Soviet Central Asia, 1919-1929* (Princeton University Press 1974)

James N. Rosenau, *Citizenship Between Elections: An Inquiry Into The Mobilizable American* (Free Press 1974)

Ervin Laszio, *A Strategy For the Future: The Systems Approach to World Order* (Braziller 1974)

John R. Vincent, *Nonintervention and International Order* (Princeton University Press 1974)

Jan H. Kalicki, *The Pattern of Sino-American Crises: Political-Military Interactions in the 1950s* (Cambridge University Press 1975)

Klaus Knorr, *The Power of Nations: The Political Economy of International Relations* (Basic Books 1975)

James P. Sewell, *UNESCO and World Politics: Engaging in International Relations* (Princeton University Press 1975)

Richard A. Falk, *A Global Approach to National Policy* (Harvard University Press 1975)

Harry Eckstein and Ted Robert Gurr, *Patterns of Authority: A Structural Basis for Political Inquiry* (John Wiley & Sons 1975)

Cyril E. Black, Marius B. Jansen, Herbert S. Levine, Marion J. Levy, Jr., Henry Rosovsky, Gilbert Rozman, Henry D. Smith, II, and S. Frederick Starr, *The Modernization of Japan and Russia* (Free Press 1975)

Leon Gordenker, *International Aid and National Decisions: Development Programs in Malawi, Tanzania, and Zambia* (Princeton University Press 1976)

Carl Von Clausewitz, *On War*, edited and translated by Michael Howard and Peter Paret (Princeton University Press 1976)

Gerald Garvey and Lou Ann Garvey, eds., *International Resource Flows* (Lexington Books, D.C. Heath 1977)